Rangeland Ecosystems in the Himalayan Mountains

Rangeland Ecosystems in the Himalayan Mountains

By
Vir Singh
R.D. Gaur

2008
DAYA PUBLISHING HOUSE
Delhi - 110 035

ISBN10 81-7035-534-6
ISBN13 978-81-7035-534-2

Published by : **Daya Publishing House**
1123/74, Deva Ram Park
Tri Nagar, Delhi - 110 035
Phone: 27383999
Fax: (011) 23260116
e-mail : dayabooks@vsnl.com
website : www.dayabooks.com

Showroom : 4760-61/23, Ansari Road, Darya Ganj,
New Delhi - 110 002
Phone: 23245578, 23244987

Laser Typesetting : **Classic Computer Services**
Delhi - 110 035

Printed at : **Chawla Offset Printers**
Delhi - 110 052

PRINTED IN INDIA

Preface

Rangelands comprise the largest chunk of the geographical area in the Himalayan mountains. Nevertheless these ecosystems are the neglected ones and are often missed in the mountain development interventions. The rangelands are critical to the very ecological integrity of the mountain agro-ecosystems and are also critical to the livelihoods of millions of people inhabiting the mountains.

Looking across the scant rangeland studies carried out in the past, which are crucial for foreseeing the very future of the ecosystems of vital significance for the mountains as well as for the Earth as a whole, the book highlights virtually all issues ranging from the phytosociological characteristics to ecological and economic attributes of the range resources.

Mountain communities are largely livestock-dependent. Dependence of a large population of livestock on rangelands primarily for fodder supplies is of critical importance for the mountains' farming communities. A taxonomic account of the plant species of certain fodder value forms the major content of the book. Rangelands also harbour magnificent wildlife. A comprehensive picture of the Himalayan fauna has also been presented. Numerous ecological attributes and economic contributions of the range ecosystems as well as the various natural and anthropogenic factors and constraints operating on these ecosystems have been discussed at length.

This book is a product of our studies focused on various issues of the rangelands in the Himalayan mountains. Though the main body of the content is based on the studies of various types of rangelands in the Uttarakhand Himalaya, also referred to as the Indian Central Himalaya, it would also reflect the scenarios of the entire Himalayan region and be of help in addressing the issues pertaining to this crucial geographical region of the globe.

The Department of Botany of the HNB Garhwal University provided required help and environment conducive to the study which has been translated into this book. The Department of Environmental Sciences of the GB Pant University of Agriculture and Technology also helped in creating stimulating environment for completing the project of book writing.

Many of our friends have been instrumental in the accomplishment of this task. We are extremely grateful to Professor Y.P.S. Pangtey, formerly with the Botany Department of the Kumaun University and currently an advisor to the Government of Uttarakhand, for his kind cooperation. Special thanks are also due to Prof. C.M. Sharma of the Botany Department, HNB Garhwal University; Dr. K.S. Negi, In-charge, National Bureau of Plant Genetic Resources, Bhowali Centre; Dr. G.C.S. Negi of the GB Pant Institute of Himalayan Environment and Development, Almora.

Prof. PL Gautam, Vice-Chancellor of the GB Pant University of Agriculture and Technology has motivated many minds to take up interest in the rangeland ecosystems and taking keen interest in present day ecological problems. Many scientists at the G.B. Pant University of Agriculture and Technology and Botany Department of HNB Garhwal University have been of kind help during this study. Thanks are due to Dr. Ashok Kumar, Professor of Animal Nutrition and In-charge University Fodder Bank for providing commercial feed; to Dr. R.S. Jaiswal, Associate Professor for help in lab analysis of the fodder samples; to Er. H.J. Shiva Prasad, Associate Professor of the College of Technology for providing facilities for the use of electronic equipment in his office and to Dr. Govind Singh Kushwaha, Assistant Professor of Psychology for moral support.

Facilities extended by the G.B. Pant University of Agriculture and Technology, especially by Prof. G.K. Garg and Prof. D.P. Mishra, the former Deans of the College of Basic Sciences and Humanities and Prof. Uma Melkania, Head, Department of Environmental Sciences are gratefully acknowledged.

We would like to appreciate the efforts of Dr. Babita Bohra, presently a Rangeland Consultant at ICIMOD, Kathmandu, Nepal, who has provided much needed help particularly in the preparation of Herbarium and chemical and statistical analyses of the fodder samples. We would also like to thank the villagers of the Almora and Champawat districts who provided invaluable help during field study. One the authors (R.D. Gaur) is thankful to UGC-New Delhi for providing Emeritus Fellowship.

We must also appreciate the trust, moral inputs and forbearance of our families, which have been the pivotal ingredients in the accomplishment of this project.

This book could be a highly useful tool for the students, teachers, researchers and natural resource management experts and to all those interested in botany, ecology, environmental sciences, wildlife, animal sciences, sociology and many other aspects of mountains' cultures. We earnestly hope that the book would be of some value towards stirring the conscience of development proponents, policy makers and R&D systems, which might have implications for the protection, conservation, regeneration and development of the range ecosystems in the fragile Himalaya. We also hope that the book would contribute to the promotion of rangeland ecology for building up a strong base for sustainable development of the Himalayan region.

Vir Singh

R.D. Gaur

facilities extended by the G.B. Pant University of Agriculture and Technology especially by Prof. C.K. Garg and Prof. D.P. Mishra, the former Dean of the College of Basic Sciences and Humanities and Prof. Uma Melkania, Head, Department of Environmental Sciences are gratefully acknowledged.

We would like to appreciate the interest of Dr. [illegible] Bohra, presently Rangeland Consultant at ICIMOD Kathmandu, Nepal who has provided much needed help particularly in the preparation of illustration and chemical and statistical analyses of the fodder samples. We would also like to thank the villagers of the Almora and Champawat districts who provided invaluable help during the field study. One the author (R.D. Gaur) is thankful to UGC New Delhi for providing Emeritus Fellowship.

We must also appreciate the moral input and forbearance of our families, which have been the pivotal ingredients in the accomplishment of this project.

This book could be a highly useful tool for the students, teachers, researchers and natural resource management experts and to all those interested in pasture ecology, environmental sciences, wildlife, animal science, ecology and many other aspects of mountain challenges. We earnestly hope that the book would be of some value towards ensuring the sustenance of development components like grazing lands and the ecosystems, which might have implications for the protection, conservation, recovery and development of the range ecosystems in the fragile Himalaya. We also hope that the book would contribute to the promotion of range-lands ecology for building up a strong base for sustainable development of the Himalayan region.

Vir Singh

R.D. Gaur

Contents

Acronyms and Abbreviations

%	Per cent
°C	Degree centigrade
A	Abundance
ADF	Acid detergent fibre
AIA	Acid insoluble ash
BSD	Herbarium, Botanical Survey India, Dehradun
CF	Crude fibre
CFB	Complete feed block
cm	Centimeter
CP	Crude protein
CPR	Common property resources
CRD	Complete Randomised Design
DAP	Draught animal power

DM	Dry matter
DMD	Dry matter digestibility
e.g.	*exampli gratis*; for example
EE	Ether extract
et al.	*et alia*; and others
etc.	*et cetera;* and others
F	Frequency
Fig.	*Figura*; figure, illustration
g	Gram
GUH	Herbarium, HNB Garhwal University
ha	Hectare
HC	Hemicelluloses
HKH	Hindu Kush-Himalaya
i.e.	*id est*; that is
IK	Indigenous knowledge
IVI	Importance Value Index
J. and K.	Jammu and Kashmir
m	Metre
mm	Millimetre
msl	Mean sea level
NDF	Neutral detergent fibre
NFE	Nitrogen-free extract
NPK	Nitrogen, phosphorus and potassium
OM	Organic matter
PPR	Private property resource
sp.	*Species*; species
TA	Total ash
UMMB	Urea-molasses-mineral block
viz.	*videlicet*; namely

List of Figures

List of Tables

Chapter 1
Introduction

Himalaya, one of the richest hot spots of biodiversity of the globe, offers immense opportunities in various fields of biological domains and associated patterns of sustainable life support systems. A variety of ecosystems is flourishing in the Himalaya, but not all of them have been properly addressed by earlier workers for the want of their management. Despite the advent of various disciplines and policies with focus on conservation and sustainable development, studies on rangeland management received little attention from the vast Himalayan region, from far west J and K to easternmost of Arunachal Pradesh. Comprising the youngest, the highest and the most fragile mountain ecosystems, the Himalaya has vital impact on the life systems of South Asia and has direct bearing on the socioeconomic scenarios, particularly of the eight Asian countries, *viz.*, Afghanistan, Pakistan, China, India, Nepal, Bhutan, Myanmar and Bangladesh.

Rangelands, according to the Society of Rangeland Management Website, are the lands on which the native vegetation–predominantly grasses and grass like plants, forage or shrubs–is suitable for use by grazing and browsing. Earlier Stoddart *et al.* (1975) described rangelands as those areas of the world which by reason of physical limitations, low and erratic precipitation, rough topography, poor drainage and extreme temperatures are unsuited to cultivation and which are a source of forage for free ranging native and domestic animals, as well as a source of wood products, water and wildlife.

There are several other loose definitions of the rangeland ecosystems. In the Himalayan context, the following holistic definition is proposed herewith: *Rangelands are the uncultivated areas or abandoned cultivated areas and occasionally fallow lands that harbour natural or seeded/planted vegetation of herbaceous and woody species and that serve as habitats of a variety of wild animals, support domesticated animals, provide unique products and ecosystem services especially vital for the livelihoods of livestock-dependent communities.*

These lands encompass approximately half of the earth's land surface (Sidahmed and Rota 2004). Some estimates yet place 70 per cent of the world's total areas as rangelands (Zhaoli 2004). More than 60 per cent of the Himalayan region consists of rangeland ecosystems. These rangelands support the livestock industry, accommodate important watershed functions, and provide valuable as well as biologically diverse resources.

Rangelands support a variety of herbs, shrubs and trees, many of which are of great economic importance. At mid-altitudes in the mountains, rangelands are the result of either burning or continuous extraction of fuel, fodder and timber from the forest ecosystems at the rates exceeding the carrying capacity of these ecosystems. Rangelands at high altitude in the Greater Himalaya are the natural ones composed of the climatic climax vegetation. These alpine pastures are locally known as 'Bugyals'. These areas under the natural rangelands are covered with thick layer of snow for about six months in a year. All these pastures constitute about four per cent of the geographical area in the Uttarakhand Himalaya. These are utilised by migratory animals of the nomadic communities as well as by animals of inhabitants of the adjacent areas. Inhabited areas of the Greater Himalaya are invariably the livestock–dominated farming systems, where livelihoods of people revolve round the livestock resources feeding on the rangelands.

Rangelands constitute the core component of mountain farming systems. These ecosystems not only render crucial ecological services even to far off forelands, but also contribute to the livelihood security of livestock dependent–communities/ pastoralists dominating the region. Diversification of economic base is the main opportunity for the marginal farmers dominating in mountain habitats of the Himalaya to increase their income and create an environment of economic security. The rangeland ecosystems carry many opportunities for diversified land-based activities leading to socio-economic upliftment of the local people. The rangeland ecosystems are not only the rich repositories of highly nutritive forage for livestock, but are also the natural home to a large number of rare medicinal, aromatic, and ornamental plants. Adapted to local environmental stresses, several less understood and so far unexploited or under-exploited plant species of mountain rangelands have substantial economic value for the local inhabitants. Rangelands of high Himalaya are also suitable for cultivation of some of the high-value food grains, *e.g.*, amaranth, buckwheat, seed chenopod and naked barley, as well as a variety of kidney beans.

Apart from being the ecosystems of vital ecological attributes to the mountains as well as to the foreland areas, the rangelands constitute major component of the livestock-based economies in the Himalayan region. Surprisingly, however, rangelands are the most neglected ecosystems in the mountains. Their vital ecological services, socio-cultural and economic attributes are not appreciated as much as these uncultivated areas deserve. These seldom occupy significant place in the institutional policies, plans and development programmes. The entire Himalayan region is in the throes of vicious ecological and livelihood crises, obviously, to a large extent, on account of the neglect of the management of the rangeland ecosystems.

Harbouring a unique diversity of flora and fauna, the rangelands are an integral part of mountain farming systems. An investigation into the current status of the rangelands in the Himalaya, especially in terms of their floristic composition, herbage productivity, wildlife diversity, factors and processes responsible for their degradation/ regeneration, nutritive value of the fodder species flourishing in their environment, linkages with other farming system components in terms of biomass/ energy flows, etc. would be of great significance towards a better understanding of these less addressed ecosystems of the Himalayan region.

Local livestock-dependent communities, contrary to the institutional systems, give very high value to these ecosystems. They do have their own strategies of rangeland management that have evolved and are evolving through trials and errors experiences. Traditional knowledge relating to rangeland management is crucial for evolving sound livelihood systems and, therefore, needs to be documented, evaluated, demystified and articulated in our modern approaches to natural resource management in the mountain areas. Mountain rangelands have a cultural value for local population. Farming communities have customary traditional practices to maintain rangelands as a sustainable resource system. Simultaneously, biodiversity management in an ecosystem represents one of the wonderful outcomes of traditional knowledge. Farmers are cautious to maintain diversity of woody perennials and herbaceous plants. Prospering biodiversity and its management in the Himalaya need to be taken stock of for sustainable rangelands. Large proportion of the biodiversity occurring in these ecosystems is utilized to maintain animal husbandry in the region. How this biodiversity is utilized is also one of the important issues that need to be addressed.

Himalayan rangelands are the flourishing case of common property resources (CPRs) and their management is a peculiar case of community-based management. An understanding of the community-based knowledge and technical skills relating to resource management will be the foremost guiding factor that could be articulated in mitigating local conflicts and issues along with giving way to an ecologically sound management of this precious resource system.

Rangelands, croplands and livestock in the Himalayan mountains are integrated with considerable nutrient flows within a single functional unit, the agro-ecosystem. Amongst these, rangelands serve as a storehouse of nutrients. Nutrients contained in the vegetation flow into croplands via livestock. The livestock use plant nutrients for their maintenance and production. At the same time they help fast degradation of the rangeland biomass with the help of rumen microorganisms. The manure thus produced is transferred into the croplands. The cropland biomass (crop residues) is also consumed by livestock and in the same way a proportion of nutrients is recycled into the cropland. Yet a proportion of the nutrients is recycled into the rangelands through dung voided by grazing animals. Such nutrient cycling keeps a mountain agro-ecosystem in a state of dynamic equilibrium and makes it a self-sustaining unit of nature. Rangelands play a key role towards maintaining a state of dynamic equilibrium and self-sustainability.

Voluminous literature on various issues of rangelands available so far is not rich enough to be dependent upon for the ecological regeneration and development of these ecosystems. While some appreciable research work has been conducted on the mountain rangelands in the Hindu Kush-Himalayan (HKH) Region (*e.g.*, Alirol 1979; Pandey 1982; Amatya 1990; Zhaoli 2004, etc.), Uttarakhand Himalaya appears to be lagging behind in this regard. Some of the aspects of rangelands in Uttarakhand have been covered in the studies by GB Pant University (1982), Singh (1985, 1989, 2002a), Singh and Naik (1987), Singh *et al.* (1995), Solanki *et al.* (1999), Bana and Singh (2004), Gaur (2004), Singh and Bohra (2005), Singh and Gaur (2005) and Bohra (2006).

These studies included only some scattered information on the rangeland/grassland ecosystems and fodder resources in Uttarakhand Himalaya. Broadly, the whole concept of rangelands, particularly of Uttarakhand Himalaya, suffers from the information gap, systematic studies and lacks complimentary aspects. Attempts have been made in the subsequent Chapters in this book to bridge up this gap. A deeper understanding of the ecology, management, socioeconomic attributes and many other environmental services of the rangeland ecosystems in the Himalayan mountains is of absolute importance for the sustainable development of livelihood systems in the region.

Chapter 2

Rangelands in Retrospect

Mountain farming systems comprise considerably large area under forests and rangelands. A number of forest types occur in the Himalaya, which are distributed widely with the gradient of elevation, ranging from sub-tropical to temperate and alpine zones (Champion and Seth 1968; Singh and Singh 1992). Livestock are the main component of the mixed farming systems in the Himalayan areas. They convert the energy of forests, grasslands and crop residues into useful products, like draught power, milk, meat, wool, manure, etc. and thus play a crucial role in livelihood systems. A number of comprehensive studies have been conducted on fodder resources and livestock production systems in the mountain areas (GB Pant University 1982; Jackson 1985; Singh 1985, 1989, 2000, 2002b, 2005; Jodha and Shrestha 1990; Singh and Gaur 2005; Singh and Bohra 2005; Bohra 2006). The livestock provide tangible and intangible benefits to the mountain farmers (Singh 1998a). Smallholder dairy farming is increasingly gaining importance as a source of family income in mountain areas for quite some time (Singh 1999, 2000, 2001, 2002b; Singh and Vaidya 2002; Singh *et al.*, 2001a, 2001b, 2001c; Tulachan *et al.*, 2002).

Comprehensive rangeland studies in Uttarakhand mountains are scant. Studies by Singh and Saxena (1978), Gaur (1982, 1999, 2004), GB Pant University (1982), Singh (1985, 1989, 1991, 1993, 1994, 2000, 2002a, 2002b, 2004a, 2004b), Singh and Naik (1987), Debnath and Singh (1990), Gaur and Maithani (1990), Gaur *et al.* (1983, 1985), Misri (1986), Negi and Gaur (1994a, 1994b), Negi *et al.* (1985, 1995), Rawat *et al.* (1993), Misri (1999), Singh *et al.* (1995, 1998, 1999, 2001a, 2001b, 2001c, 2002, 2003), Solanki *et al.* (1999), Singh and Kumar (2002), Singh and Tripathi (2003), Yadava *et al.* (2003), Bana and Singh (2004), Chaturvedi *et al.* (2004), Joshi and Singh (2004), Singh and Gaur (2005), Singh and Bohra (2005), and Bohra (2006) have incorporated some components of rangelands in Uttarakhand. Of late, American Society of Range Management (1964), Vallentine (1971) and Stoddart *et al.* (1975) focused on the general aspects of rangelands; Palmieri (1976), Alirol (1979), Pandey (1982), Brower (1987), Amatya (1990), Misri (1995), Pariyar *et al.* (1996), Zhaoli (2004) and many other workers, particularly at the International Centre for Integrated Mountain Development (ICIMOD), Kathmandu have done appreciable research work on the rangelands of different parts of the HKH region.

The Rangelands

Grasslands or grazing lands are often used as synonym of rangelands. Though these terms can be used interchangeably, the latter carry a wider dimension in terms of their floristic composition and utilisation. Rangelands can be dominated by diverse vegetation and are often associated with livelihood systems of the communities who manage or have access to them.

In India it seems very doubtful if there are any examples of tropical climax grassland, though grassland is common and it may be very stable preclimax under the influence of fire and grazing (Champion 1936). Only the Himalayan alpine meadows have the status of climatic climax grasslands in India. R. Mishra, in his presidential address to the section of botany at the Forty-sixth Session of the Indian Science Congress in 1959, had reminded the ecologists that apart from the high Himalayan alpine meadows there is no climatic climax of grassland in India (Yadav and Singh 1977). Monsoon grasslands occurring in a forest climate could be called tertiary communities to distinguish them from secondary communities constituted by shrub and shrub-tree complexes (Whyte 1974). Supporting this, Yadav and Singh (1977) state logically, "Because of their seral nature, the grassland communities change their composition when the grazing or burning pressure is relaxed. Complete, long-term protection often moves the grassland towards a forest climax."

Rangeland management is the science and art of optimizing the return from rangeland through the manipulation of range ecosystems (Stoddart *et al.*, 1975). In 1973, National Committee on Rangeland Management defined rangelands as "uncultivated areas (although sometimes disturbed by unthoughtful cultivation) that support natural or seeded herbaceous or shrubby vegetation with or without trees". The American Society for Range Management (1964) defined range management as the art and science of planning and directing range use to obtain sustained maximum animal production consistent with perpetuation of natural resources. Viewed within the system framework, rangeland management has been defined as management of renewable natural resource composed of one or more range ecosystems for the optimum sustained yield by the combination of goods and services (Vallentine 1971).

Rangelands are major ecosystems within the HKH region and, for a large number of local people, they provide the main or the only livelihood resource. At the same time, these rangelands also provide high value ecosystem functions and services (Zhaoli 2004). Gaur (1999) has brought out comprehensive account of the biodiversity present in the Garhwal district of Uttarakhand, which also gives an account of the biodiversity occurring in the rangelands of other Himalayan areas.

Many studies suggest that dismal performance of livestock in the Central Himalaya is mainly attributable to the scarcity of the fodder. Common property resources (CPRs), which are parts of the rangelands, rather than cropland area, fulfill most of the livestock fodder requirements in Central Himalaya as well as in other areas of the Himalaya (Singh and Naik 1987; Singh 1996, 2001; Singh and Bohra 2005). A comprehensive account of fodder trees of India, including those found in grassland ecosystems in the mountains has been provided by Singh (1982). Singh and Naik (1987), Singh (1989) and Singh *et al.* (1995) have presented their perspective on the utilization of forest based fodder in the mountains, while Singh *et al.*

(1988, 1990) have analysed some of the fodders extracted from range trees. Pandey (1982) and Amatya (1990) have presented details on fodder trees and lopping cycle in the mountain areas of Nepal.

Biodiversity is influenced considerably by altitude and slope orientation. Grasslands at lower and mid-altitudes are often interspersed by trees and shrubs. People's management gives performance to the fodder-yielding species. Amatya (1990), Bhatt and Rawat (1993), Bana *et al*. (1996), Bana and Singh (2004), Singh and Gaur (2005), Singh and Bohra (2005) and Bohra (2006) have identified and documented an impressively large number of fodder species occurring in the rangeland ecosystems in parts of HKH region. Prevalence of individual species is mainly according to altitude and slope orientation.

Livestock feeding management is optimized on the basis of lopping cycle of individual species. Phenology gives an idea of the lopping cycle of individual fodder species and, in essence, is a useful tool for animal scientists and farmers in the mountain areas. It contributes to increase production and develop livestock based enterprises and consequently to socio-economic development of livestock dependent farming community in the region (Bana and Singh 2004).

Rangelands being the core components of the farming systems in the HKH region hold key to ecological integrity, a vital prerequisite of agricultural sustainability (Singh 1995, 1998b). Rangelands constitute the major area in the mountain farming systems. Feed and fodder occurring in those ecosystems are utilized by the livestock. Nutrients contained in the rangelands thus flow into the crop lands via manure (Singh and Bohra 2005). Mountain areas comprise crop-livestock mixed farming systems. The farmers in these areas practice pastoralism. They comprise large size of herds, especially the goats and the sheep (Singh 1998a).

Diversity of Fodder Species

The favoured fodder zone lies between 2000m altitude and snow line in Jammu and Kashmir, Swat, Kaghan valleys, and northern areas. These areas provide excellent forage for livestock grazing during summer growing seasons (Khan 1979). There is very high degree of diversity of fodder species in the mountains. The favoured fodder trees are *Acacia, Artocarpus, Albizia, Bauhinia,*

Dalbergia, Dendrocalamus, Ficus, Grewia, Terminalia, and *Toona* (Misri 1999).

Temperate pasture lands are associated with oak or mixed broad-leaved species such as *Quercus* or blue pine forests. The common forage species are *Arundinella hookerii, Andropogon tristis, Poa* spp., *Chrysopogon gryllus, Dactylis glomerata, Stipa concinna, Festuca* spp., *Cymbopogon* spp., *Bothriochloa* spp., *Desmodium* spp. and *Agrostis micrantha.* Provision of feed and fodder to livestock in mountain areas would largely depend on the type of ecological zones Pariyar *et al.* (1996) (Table 2.1).

Several studies (*e.g.*, GB Pant University 1982; Singh 1985; Singh and Naik 1987; Singh *et al.*, 1988; Singh 2001; Singh and Bohra 2005; Bohra 2006) suggest that the dismal performance of livestock in Uttarakhand Himalaya is mainly attributed to the scarcity of fodder. Rangelands and other fodder resources in the Himalaya have shrunken due to a number of factors over last few years (Singh 1992, Singh 1998b). In most of the cases, the rangelands are the part of the common property resources (CPRs) and fulfill most of the livestock fodder requirement, far much more than the cropland area (Singh and Naik 1987; Singh 2000, 2001; Singh and Bohra). Rangelands' diversity of vegetation also supports variety of wild animals. Gaur (1999) describes vegetation and peculiar environmental factors that sustain a variety of wildlife in parts of the Himalaya.

Roder *et al.* (2001) have presented comprehensive matter on the feeding of livestock and improvement of fodder resources in Bhutan. The matter, nevertheless, has to have bearing on all other areas of the Himalayan Mountains.

All these studies would suggest that the uncultivated component of the mountain farming systems–*i.e.*, the rangelands and the forests–have key role to play in the sustenance of dairy farming and are the only key to augment dairy production in mountain areas. Efforts have been made to quantify the role of the rangeland ecosystems in animal production systems (GB Pant University 1982; Jackson 1985; Singh 1985, 2000; Singh and Naik 1987; Singh and Sharma 1990; Tripathi 2001; Bohra 2003, 2006; Bohra and Singh 2005). These studies also suggest impact of the available feed resources to be derived from the rangeland / grassland ecosystems on various parameters of dairy production.

Table 2.1: Potential Forage Species for Different Agro-ecological Regions

High-hills	*Mid-hills*	*Terai*
Trifolium repens cv. Redquin	*Trifolium alexandrinum*	*Trifolium alexandrinum*
Dactylis glomerata cv. Porto	*Avena sativa*	*Avena sativa*
Medicago sativa cv. Southern special	*Pennisetium purpureum*	*Vicia faba*
Medicago sativa cv. Hunter river	*Setaria anceps*	*Euchlaena mexicana*
Agropyron intermedium cv. *Dhae*	*Desmodium intortum*	*Sorghum bicolor*
Agropyron elogatum cv. Camison	*Melinis minutiflora*	*Vicia sativa*
Trifolium repens cv. Khumal	*Stylosanthes*	
Trifolium hybridium	*Pueraria Phaseoloides*	
Lolium perenne cv. *Prior*		
Dactylis glomerata cv.		
Tammisto		
Trifolium repens cv. *184*		
Trifolium repens cv.		
Menna		
Trifolium repens cv. *Huia*		
Lolium perenne cv. *Tyson*		
D. glomerata cv. *Cambria*		
F. arundinaceae cv. *Dovey*		
Lolium perenne cv. *Augusta*		
L. glomerata cv. *Apanium*		
T. repens cv. *Mewna*		
L. perenne + T. repens		
P. pratense cv. *Florie*		
Trifolium repens cv. *Huia*		
T. repens cv. *Ladino*		
T. pratense cv. *Pawera*		
T. pratense cv. *Broad leaf*		

Source: Pariyar *et al.* (1996).

The Trees

In many parts of India tree leaves are overwhelmingly used as source of fodder. A list of such trees occurring throughout India has been provided by Anonymous (2001) (Table 2.2).

Table 2.2: Important Species for Fodder

Grewia elastica	*Garuga pinnata*
Gmelina arborea	*Grewia optiva*
Acacia farnesiana	*Grewia tiliaefolia*
Acacia hirsutus	*Helicteris isora*
Ficus rumphii	*Heloptelia integrifolia*
Hardwickia binata	*Hymenodictyon excelsum*
Flacourtia indica	*Kydia calycina*
Albizia lebbeck	*Lagerstroemia parviflora*
Acacia decurrens	*Lagerstroemia parviflora*
Acacia lebbeck	*Lagerstroemia speciosa*
Acacia mearensii	*Lannea coramandelica*
Acacia melanoxylon	*Leucaena leucocephala*
Acacia lecophloea	*Madhuca longifolia*
Acacia modesta	*Maliotus philippensis*
Acacia nilotica	*Mangifera indica*
Acacia procera	*Manikara hexandra*
Acacia senegal	*Melia azedarach*
Acacia tortillas	*Millettia extensa*
Adina cordifolia	*Mitragyna parviflora*
Acacia dealbata	*Morinda pubescens*
Aegle marmelos	*Moringa oleifera*
Ailanthus excelsa	*Morus alba*
Albizia amara	*Morus macroura*
Albizia chinensis	*Morus serrata*
Anogeissus latifolia	*Olea ferruginea*
Acacia odoratissima	*Ougeinia oojeinensis*
Acacia pendula	*Phoenix dactylifera*
Anthocephalus chinensis	*Pithecellobium dulce*
Artocaarpus heteropphyllus	*Pongamia pinnata*
Azadirachta indica	*Populus ciliata*

Contd...

Table 2.2–Contd...

Bauhinia malabarica	*Prosopis chinensis*
Bauhinia semla	*Prosopis cineraria*
Bauhinia variegata	*Prosopis specigera*
Bombax ceiba	*Pterocarpus marsupium*
Bauhinia purpurea	*Quercus floribunda*
Artocarpus lakucha	*Quercus glauca*
Acacia catechu	*Quercus leucotrichophora*
Boswellia serrata	*Quercus semecarpifolia*
Careya arborea	*Robinia pseudo-acacia*
Bridelia retusa	*Salix tetrasperma*
Buchanania latifolia	*Saurauia napaulensis*
Butea monosperma	*Schleichera oleosa*
Celtis australis	*Sesbania grandiflora*
Cordia dichotoma	*Shorea robusta*
Dalbergia sissoo	*Soymida febrifuga*
Delonix regia	*Spondias pinnata*
Dillenia pentagyna	*Stereospermum chelonoides*
Diospyros melanoxylon	*Syzygium cuminii*
Diospyros tomentosa	*Tamarindus indica*
Ehertia laevis	*Terminalia alata*
Ficus lacor	*Terminalia arjuna*
Ficus hispida	*Terminalia bellerica*
Ficus sycomoros	*Terminalia chebula*
Ficus benghalensis	*Terminalia crenulata*
Ficus palmata	*Terminalia myriocarpa*
Ficus racemosa	*Terminalia paniculata*
Ficus auriculata	*Tinospora cordifolia*
Ficus religiosa	*Toona ciliata*
Ficus semicordata	*Trema orientalis*
Ficus microcarpa	*Wrightia tomentosa*
Gardenia latifolia	*Ziziphus mauritiana*
Gardinia turgida	*Ziziphus nummularia*
	Ziziphus xylopyrus

Source: Anonymous (2001).

Fodder trees can be said to be the most important and constant fodder resources in the Himalyan region. Balaraman (1981) has presented a list of 44 fodder trees from Sikkim Himalya. Many of these fodder tress occur in other parts of Indian Himalaya. Singh (1982) has given an exhaustive list virtually of all fodder trees in India, including those occurring in the Himalyan region. Singh (1985) has reported 9 fodder trees from Kumaun Himalaya which are most common for fodder for livestock. Amatya (1990) has presented a list of some 50 fodder trees from Nepal Himalya along with their lopping cycle. Bhatt and Rawat (1993) have reported 14 commonly utilized fodder trees from Uttarakhand Himalaya.

Fodder trees are of multiple use. Along with nutritive fodder during different seasons, these trees also provide a number of other useful products. Giving logic behind the use of rangeland-based uncultivated fodders, Singh *et al*. (1995) have brought to the fore the several attributes of the system based on uncultivated fodders, especially trees, *viz.*, response to fragility, management of diversity, harness of ecological niches, higher productivity, comparable nutritive value, fodder supply throughout the year, enhanced resilience, compatibility with the specific mountain features, topographic suitability, high energy efficiency, multipurposeness, environment friendliness and sustainability.

The Grasses

Grasses comprise the dominant herbaceous vegetation in a typical rangeland. Influence of grasses on man and civilisation has been incalculable but their biology is almost neglected in both graduate and postgraduate studies. Our knowledge and understanding of grass biology remains rudimentary. Plant physiology tends to favour the lofty dicotyledons and gloss over the lowly grasses almost as if the monocotyledonous plants once pigeonholed are scarcely worthy of intense study. Underlying the importance of grasses, Dugger (1972) relates that about 3000 years ago, Hebrew prophet Isaiah wrote: "All fresh is grass" and for centuries that concept fascinated philosophers, poets, and theologians who in our more agrarian-oriented past understood that the idea was not only metaphorical but literal. Further stressing the importance of the grasses, he recalls that wheat, rice, corn and other cereal grains–primary staples of mankind since long before recorded history–are all grasses. Grasses with their 10,000 species not only far surpass all other vegetation of the world in abundance, but they remain the most important economic family of plants to man.

J. Ingals has the following in the praise of grass (cited in Richard *et al.*, 2000):

> *Grass is the forgiveness of nature–her constant benediction. Fields trampled with battle, saturated with blood, torn with the ruts of cannon, grow green again with grass and carnage is forgotten. Streets abandoned by traffic become grass-grown like rural lanes, and are obliterated. Forests decay, harvests perish, flowers vanish, but grass is immortal. Beleaguered by the sullen gusts of winter, it withdraws into the impregnable fortress of the subterranean vitality and emerges upon the first solicitation of spring. Sown by the winds, by wandering birds, propagated by the subtle horticulture of elements, which are its servants and masters, it softens the rude outline of the world. It invades the solitude of the forests, climbs the inaccessible slopes and forbidding pinnacles of the mountains, modifies climates and determines history, character and destiny of nations.*

Puri (1960) classified the grassland communities into three broad categories on the basis of moisture, which have been elaborated by Yadav and Singh (1977) as follows:

Xerophilous Grasslands

Covering the regions of northwest India, these occur in semi-desert conditions. *Eragrostis* is the dominant grass genus and other common grasses associated with it are *Aristida adscensionis*, *Heteropogon contortus*, *Erimopogon foveolatus*, *Chloris virgata*, *Apluda varia*, *Cenchrus ciliaris*, *C. barbatus*, and *Lasiurus hirsutus*. *Prosopis cineraria*, *Capparis* sp., and *Balanites* sp. are the usual ligneous species occurring in these grasslands.

Mesophilous Grasslands

Saccharum munja, *S. narenga*, and *Vetiveria zizanioides* are predominant grasses, spreading in the region of typical moist deciduous forests of Uttar Pradesh. These grasslands may give the appearance of a savanna. Alpine meadows of Himalaya and the grassy downs of the Nilgiris are temperate variations of this type.

Hygrophilous Grasslands

These give the appearance of wet savanna, dominated by *Erianthus elephantinus* and *Saccharum arundinaceum*. *Erianthus ravennae* and *Phragmites-Arundo* stands occur on drier and wetter sites, respectively.

Illustrating the evolution of grass family and supporting the treatments of angiosperm phylogeny of Hutchinson (1973), Takhtajan (1981), Cronquist (1981) and Thorne (1968), Stebbins (1972) agrees in placing the family at one of the most evolved taxa of the evolutionary "tree" of flowering plants. The apparent simplicity of both their vegetative and reproductive structures results from phylogenetic reduction and is a sign of extreme specialization. The similarity between grasses and other large family of grasslike plants, the sedges (Cyperaceae), is only superficial and is no indication of a true relationship. Sedges differ radically from grasses in the structure of their spikelets and flowers as well as in their chromosomes, which have the peculiar condition of multiple centromeres (Tanaka 1949; Ostergren 1949). In all of these features Cyperaceae are more like Juncaceae than grasses. These two families, as further suggested by Stebbins (1972) are probably derived from a common ancestor.

Development of range grasses for higher forage yields under natural competition with other species requires the preservation of greater adaptability than that needed by crops grown in pure culture under the artificial environment of cultivation. Describing polyploidy as a factor in the evolution and distribution of grasses, Johnson (1972) related that the versatility of polyploids, indicated by their striking role in the evolution of The *Gramineae*, suggests that induced amphiploidy in particular has a place among techniques for the breeding of such grasses.

Origin of our major grasses is an important factor in any consideration of grasses and their improvement through breeding. Most of the grasses used for hay pasture, and turf, and many species adapted for range seedling in arid and semiarid regions were introduced to the United States from other continents. Hanson (1972) has synthesised the work of grass introduction from one part of the world to the other. The pattern of introduction was established soon after the settlement of the Eastern seaboard, "English grass", a collective term for mixtures of bluegrasses (*Poa* spp.), ryegrasses (*Lolium* spp.), bentgrasses (*Agrostis* spp.), fescues (*Festuca* spp.), and white clover (*Trifolium repens*) were seeded by early settlers in New England. The "seed" was obtained in the form of chaff collected from immigrant ships. In Rhode Island hay seed chaff was used to plant hay meadows as early as 1647, and by 1650 there was some movement of hay seed among the New England colonies (Colby 1941).

Bermuda grass, *Cynodon dactylon*, which was brought to Savannah, Georgia, in 1751 by Governor Harry Ellis, was considered to be one of the most important grasses in the Southern United States by 1807 (Burton 1951). This observation is substantiated in early agricultural reports from U.S. Commissioner of Patents (Anonymous 1850). Guinea grass, *Panicum maximum* was introduced as early as 1813 into Mississippi, and Johnson grass, *Sorghum halepense* was sent to South Carolina from Turkey in 1830. The accidental introduction of Dallis grass, *Paspalum dilatatum* from South America is thought to have occurred before 1850.

Transfer of exotic germplasm to a new environment provides an immediate opportunity for new ecotypes to evolve by natural selection, provided that some portion of the original introduction survives. Differences in photoperiod, temperature extremes, available moisture, soil fertility, drainage, disease organisms, and insects contribute to the origin of adapted ecotypes (Hanson 1972). The advantages of locally grown clover used were recognized as early as 1850 in the United States (Furgus and Hollowell 1960) and many field experiments have served to confirm this point (Hanson 1972).

The general goal of the plant breeder working on turf species is to modify plant genotype through appropriate breeding procedures that will better fit the plant to its intended use (Long 1972). Although the grass plant is an assemblage of shoots, the individual tiller has been considered by many investigators to be the growth unit of grasses.

Productivity depends in large measure on the ability of the plant to initiate tillers and on the later development of these shoots. Within the limits set by genotype, growth of the tiller is responsive to the environment (Laude 1972). External factors that affect growth and development, therefore, warrant consideration. Investigators studying plant response to environmental factors recognize that the effect of any one factor is conditioned by the magnitude or intensity of others (Langer 1963; Laude 1972).

In the alpine zone of the Himalaya 3200 m msl, alpine meadows thrive without any stands of woody perennials. Climate of this zone is too cold and severe to allow woody perennials to grow. Alpine meadows remain covered with snow for more than six months in a year. On the onset of summer, the pastures come under biotic pressure of grazing by goats, sheep, cattle and buffaloes.

Yadav and Singh (1977) have given the account of the alpine and sub-alpine grasslands communities and species as indicated in Table 2.3.

The temperate grasslands are characteristic of the oak-conifer forest region where the soil is deep and moist. Stable grasslands occur due to moisture of forest vegetation. Animals are allowed to graze in these grasslands during winter season also. The following are the major grassland types in the temperate zone (after Yadav and Singh 1977):

Chysopogon gryllus Type

Chysopogon gryllus is associated with *Koeleria cristata* in the upper part of the temperate region, while at lower latitudes it is associated with *Heteropogon contortus, Themeda anathera* and *Bothriochloa pertusa. Eragrostis anathera, Polypogon fugax, Rottboellia exaltata, Agrostis canina. Poa* and *Festuca* sp. are the common grasses. The *Chysopogon gryllus* community of this grassland type is replaced by *Cymbopogon stracheyi* community on burning. On overgrazing, all the communities are replaced by *Bothriochloa pertusa–Cynodon dactylon* community.

Koeleria cristata Type

This grassland type occurs with *Chrysopogon gryllus* type at lower altitudes and with *Danthonia* type at higher altitudes. *Koeleria cristata, Calamagrostis epijegos, Chrysopogon gryllus, Agrostis canina, Festuca valesiaca, Arundinella nepalensis,* and *Andropogon tristis* are the dominant grasses.

Poa pratensis Type

This exists mostly between 2500 m and 2800 m altitudes. The dominant grasses are *Poa pratensis, Poa annua,* and *P. stewardiana.*

Table 2.3: Grassland Type, Communities and Species in the Himalayan Alpine and Sub-alpine Zones

Grassland Type	*Altitude, m msl*	*Grassland Communities*	*Common Grasses*
Agrostis type	4000–5200	*Agrostis munroana–Trisetum aeneum, Helictotrichon virescens, Agrostis munroana–Poa pagophila–Poa annua*	*Agrostis canina, A. munroana, Calamogrostis epijejos, Fesuca valesiaca, Hierochloa laxa, Poa annua, P. pagophila, P. stewardiana, P. pratensis, Polypogon fugax, Stipa concinna, S. sibirica, Trisetum aeneum, T. clarkei, T. flavescens, T. spicatum and Andropogon tristis*
Danthonia type	3200–4200	*Danthonia jacquemontii–Agrostis munroana–Agrostis canina, Danthonia jacquemontii* (var. *minor*)*–Triseturn spicatum and Danthonia jacquemontii–Andropogon tristis*	*Danthonia jacquemontii* var. *minor, Danthonia jacquemontii* var. *jacquemontii, D. cachemyriana, Koeleria cristata, Calamogroitis emodensis, C. epigejos, Festuca lucida, Poa annua, P. pratensis, Brachypodium sylvaticum, Stipa sibirica, Trisetum clarkei, T. spicatum, Andropogon tristis*
Puccinellia type	Confined to the eroded slopes of alpine region of Kashmir	*Puccinellia kashmiriana–Trisetum spicatum, Poa–Puccinellia*	*Puccinellia kashmiriana, Trisetum spicatum, Poa alpinia, P. pratensis, P. annua, Heirochloa laxa.*
Phleum alpinum type	Occurs on well-drained slopes and flats of meadows with deep soils and plenty of soil moisture	*Phleum alpinum, Agropyron, Deschampsia caespitosa–Deyeuxia*	*Phleum alpinum, Agropyron semicostatum, A. canaliculatum, Poa pogophilla, P. jaunsarensis, P. alpina, P. annua, Calamogrostis littorea, C. epigejos, Agrostis munroana, Festuca ovina, F. kashmiriana, Deschampsia caespitosa, Deyeuxia scabrescens*

Source: Adapted from Yadav and Singh (1977).

Themeda anathera Type

It occurs between 1700 m and 2200 m within the oak-conifer forest area. *Themeda anathera, Arundinella setosa, A. nepalensis, Polypogon fugax, Melium effusum, Bothriochloa intermedia, Oryzopsis aequiglumis, Cynodon dactylon, Festuca gigantean, Phalaris arundinacea, Bromus ramosus,* and *Eragrostis poaeoides* are the dominant grasses of this grassland type.

Main Fodder Resources

Main fodder resources are the ones which are often talked about by the livestock farmers, such as pastures or grasslands, and are critical to their livelihood systems. These are different from the species specific or lesser known fodder resources that compose large proportion of the whole plant biodiversity. The pasture lands have historically been a major source of livestock rearing in the mountains from the several past generations. They have been utilized in a variety of ways for sedentary and migratory grazing. Forests, grasslands and agroforestry systems contain a large number of plants–trees, herbs and shrubs–that provide fodder of varying nutritive values. More than four dozens of woody perennials–trees and shrubs–occurring at various altitudes of the Himalayan mountains have been listed by Singh and Naik (1987), Amatya (1990), Bhatt and Rawat (1993), Singh *et al*. (1995) and Solanki *et al*. (1999) in the hills and mountains of the Central Himalaya (Uttarakhand and Nepal). Contributions of uncultivated land to fodder supplies in the hills are far more than that of cultivated land. More than 60 per cent of the fodder requirements of livestock in mountain areas are met by trees and forest floor vegetation (Singh and Naik 1987). According to an estimate by Sundriyal (1995), grasslands and forest grazing lands meet about 50–60 per cent of the total fodder needs in the Himalaya. A forest ecosystem and woody perennials of the agroforestry systems provide fodder throughout the year on sustained basis. The grazing of animals takes place on a variety of grasslands. The true pastures and grasslands are spread over an area of about 12.04 million ha. Other grazing lands are available under tree crops and groves (3.70million ha), on wastelands (1.50 million ha) and fallow lands (2.33 million ha). Pastures and grasslands have often resulted from degradation and destruction of forests until savnnas are formed (Mishra 1983). The true pastures as climax vegetation are found only in sub-alpine and alpine pastures in higher altitudes of the Himalaya. According to Dabadghao and Shankaranarayan (1973), there are five types of grasslands.

Sehima-Dichanthium

These are spread in Central Indian Plateau, Chota Nagpur and Aravallis covering an area of 1,740,000 km^2 and elevation ranges between 300-1200 m. There are 24 species of perennial grasses, 89 species of annual grasses and 129 dicots including 56 legumes.

Dichanthium-Cenchrus-Lasiurus

These are found in the north parts of Gujarat, Rajasthan, Aravalli ranges, South Western Uttar Pradesh, Punjab. They are spread over 436,000 km^2, elevation between 150-300m. There are 11 perennial grass species, 43 annual grass species and 45 dicots including 19 legumes.

Phragmites-Saccharum-Imperata

They cover an area of 2,800,000 km^2 in the Gangetic plains and the Brahamputra valley ranges between 350-1200m.

Themeda-Arundinella

These grasslands cover over 230,400 km^2 and include the states of Manipur, Assam, West Bengal, Uttarakhand, Himachal Pradesh, and Jammu and Kashmir. The elevation ranges between 350-1200m and there are 37 major perennial grasses, 32 annual grasses and 34 dicots with nine legumes.

Temperate-alpine

These are spread across altitude higher than 2100m and include the temperate and cold arid of Jammu and Kashmir, Himachal Pradesh, Uttarakhand, West Bengal and North Eastern states.

In the states of Arunachal Pradesh and Nagaland too, fodder is not cultivated (Singh and Misri 1993). Nepal pasturelands vary from subtropical grasslands at lower elevations to alpine meadows on ridge tops in the mid hills and high mountain valleys of the inner Himalayan range. Pasture lands are mostly located in the mountains and only 4 per cent pasture area is found in low-lying areas (Table 2.4).

Table 2.4: Pasture Land by Physiographic Regions in the Himalaya

Physiographic Region	Total Land Area		Pasture Land Area		
	Ha ('000,000)	%	ha ('000)	% Total Land	% of Grazing Land
Tarai	2.1	14.4	49.7	0.3	2.9
Siwaliks	1.9	12.7	20.6	0.1	1.2
Middle mountains	4.4	29.5	292.8	2.0	17.2
High mountains	2.9	19.7	507.1	3.4	29.8
High Himal	3.5	23.7	831.5	5.6	48.9
Total	**14.8**	**100.0**	**1701.7**	**11.4**	**100**

Source: Anonymous (1986).

Shankar and Gupta (1992) have classified the Indian grazing lands as fragile eco-systems and have ranked them as class IV and V in their land capability classification. The carrying capacity of these areas is 0.20 to 1.47 adult cattle units (ACU)/ ha, but the present stocking rates are much higher.

The situation of Himalayan pastures is even more alarming due to the severe pressure of the sedentary, semi-migratory and migratory grazers. Overgrazing has caused the near complete loss of edible species. Weeds such as *Stipa, Sambucus, Aconitum, Cimcifuga, Adonis* and *Sibbaldia* have heavily infested these pastures (Misri, 1995). Fodder cultivation has remained static at 4 per cent of the total cultivated areas.

Availability is another warning factor. The annual requirement for 6.9 M ha under fodder cultivation, and for improvement of an additional 1 M ha of wastelands, is 10 M tones of cultivated fodder seed, 25,000 tones of range grass and legume seed and 500 tones of fodder tree seed per year. Against this, availability is only 20, 15 and 10 per cent , respectively of these crops (Singh and Hazra 1995a).

Sundriyal (1995) has reviewed a comprehensive analysis of the ground flora in the grasslands of Garhwal. He recorded total number of species, mean number of species, mean relative cover, mean relative frequency and mean relative density of the grass species in 13 types of grasslands in Garhwal Himalaya (Table 2.5).

Table 2.5: Grassland Types, Altitudinal Range and Number of Species in Some of the Grasslands of Garhwal Himalaya (cover, frequency and density values are for dominant species only)

Grassland Types	*Elevation, m*	*Total No. of Species*	*Mean No. of Species*	*Mean Relative Cover*	*Mean Relative Frequency*	*Mean Relative Density*
Arundinella nepalensis–Eulaliopsis binata	600–900	30	12	42	75	34
Imperata cylindrica–Chrysopogon montanus	950–1500	18	12	44	55	38
Arundinella nepalensis–Chrysopogon montanus	1400–2000	31	15	68	80	32
Imperata cylindrica	1500–1900	17	11	–	–	–
Themeda anathera–Heteropogon contortus	1600–2200	15	13	–	–	–
Heteropogon contortus–Cymbopogon distans	1500–2300	14	7	41	75	27
Chrysopogon montanus–Eulalia trispicata	1800–2400	16	5	43	60	39
Eulalia trispicata–Chysopogon fulvus	2300–2700	12	6	33	60	24
Poa spp.	2300–3300	11	4	–	–	–
Calamogrostis spp.	2800–3200	8	8	56	68	45
Danthonia cachymeriana	3000–4200	28	15	68	85	50
Agrostis stolonifera	3500–4200	12	8	36	60	40
Deschampsia–Deyeuxia	3800–4600	9	7	35	65	40

Source: Sundriyal (1995).

Range-based Livestock Production

Two types of specialized livestock production systems are emerging today. The first is in the valley areas with good access to markets. Here, a specialised cattle milk production system based on Jersey crossbred is emerging. Some areas in Himachal Pradesh practice such specialization (Singh 1997; Tulachan and Neupane 1999; Singh and Vaidya 2002). Most crucial areas for reliance on cash income through milk are those near urban areas, as they are benefited with ready market near by. In Uttarakhand, the picture is somewhat different from that of Himachal Pradesh. Here, more buffaloes are dominating the herd than the crossbred cattle (Singh and Tulachan 2002).

Pariyar *et al*. (1996) reported that in Nepal there are three types of ruminant production systems: transhumance, sedentary, and stall fed. The sedentary systems prevail at the lower altitudes of the mid altitude of the mid hills (900–1000m) and utilize all the available forage in and around the village. Transhumance system is practiced in the temperate, sub-alpine and alpine regions where the cattle, buffaloes, sheep and goats migrate from one place to the other throughout the year. Two kinds of herd are involved in full transhumance: yak-chauri and sheep-goat (Palmieri 1976, Alirol 1979, Brower 1987).

With the addition of exotic/ crossbred/ improved animal and reduction of uneconomic animal in livestock population in the country, the composition of Indian livestock has remarkably changed during the last 20 years (Tiwari and Dwivedi 1997). Small and marginal farmers contributing 62 per cent to the total milk produced yearly owned 70 per cent of livestock population (Varma 1997).

According to a study by Singh (1998a), the number of cattle per household was the largest in the middle Himalayan zone under traditional agricultural system and the smallest in the villages under transformed agriculture in the same zone. The cattle size and sex ratios are major determinants of draught animal power management in mountain farming systems.

An analysis of temporal changes in ruminant population in Uttarakhand hills between 1961 and 1999 shows an overall increase of about 15 per cent in the total ruminant population. The most noticeable change over a period of about four decades is an over whelming increase in the population of buffaloes (55 per cent) and goats (50 per cent). A considerable decline in the population of cattle (5 per cent) and sheep (14 per cent) are other noteworthy changes. With gradual emphasis on dairy sector, milk production in Uttarakhand hills has increased from mere 419 thousand MT in 1979-80 to about 715 per cent over this period. Whereas milk production from cows has increased only 19 per cent, buffalo milk production has leapt by 111 per cent over this period. The milk production increased in the Uttarakhand hills thus is largely thanks to buffaloes' contribution. Presently, buffalo population contributes more than 60 per cent of the total milk production in the region. Milk productivity per animal has increased over a period of two decades. In case of a cow it has increased from an average of 1.33 kg per day in 1979-80 to 2.30 kg per day in 1999-2000. An average increase in milk yield from 2.55 kg per day to 3.71 kg per day over the same period has been recorded for a buffalo. Lactation yield increase per cow (73 per cent) has been far higher than per buffalo (45 per cent). This is attributable to the institutional polices and programmes focusing on cows and negligence of buffaloes (Singh *et al.*, 2001a, 2001b, 2001c).

Rich body of literature on the feeding of different species of livestock in mountain areas is available. There is, in general, no particular set of standards that the hill farmers would follow. They, like their counterparts elsewhere in Indian plains, follow the thumb rules of nature while managing feeding of their livestock.

Smallholder dairy farming is becoming increasingly popular in many of the accessible pocket areas of HKH. It has also been the driving force behind the transformation of the rural economy in many parts of the Nepalese and Indian hills. An established market for fresh milk and development of milk cooperatives and milk collection centres have made this possible (Tulachan and Neupane 1999).

Describing the dairy farming characteristics in various milk sheds of the Central Himalaya, Singh (2000) has sketched the maintenance costs of dairy animals reared at various sites/ altitudes of the mountains. Singh (1997) and Singh and Vaidya (2002) have done similar studies in Himachal Pradesh Hills.

GB Pant University (1982), Jackson (1985), Singh (1985, 2000, 2002b), Singh and Sharma (1990), Tulachan and Neupane (1999),

Singh and Vaidya (2002), Tripathi (2001) and Bohra (2003, 2006) have presented strategies for livestock/ dairy development in Indian Central Himalaya. Since the key constraint to livestock production is the paucity of feed resources, augmentation of feed supplies through natural resource management and land based feed resources has been regarded as potential approach to the development of sustainable dairy farming.

There are now varying degrees of commercialization within the mixed crop-livestock farming systems. In fact, livestock production systems are becoming quite dynamic in certain pocket areas of the mountains with accessibility to road networks and market for milk. Farmers are being provided with a strong incentive to keep livestock, not just to fulfill the traditional role of providing draught power, meat, milk and manure to the households; but also to generate cash income through the sale of milk. In these areas, there has been a shift in management practices, with linkage with Common property resources (CPRs) to breakdown (Tulachan 1998, Tulachan and Neupane 1999).

In the Himalayan mountains the majority of the farmers operate in the mixed crop-livestock farming systems. There are great many different types of agro-ecosystems in this region. Geographic and topographic variations (mountain specificities–Jodha *et al.*, 1992) have combined to provide a microcosm of the earth's vegetation types and farming systems (Tulachan and Neupane 1999, Singh 2000). Owing to inherently rich resource base, such as widespread areas under rangelands, the mountain areas of Uttarakhand state, like the other similar areas, have, therefore, rightly been earmarked for dairy development. In these areas, there has been a shift in management practices, with linkage with CPRs (Tulachan and Neupane 1999).

Chemical Composition of Range Fodders

Range fodders, *i.e.*, the uncultivated fodders derivable from the leaves of certain trees or shrubs, in the context of mainstream plain areas, are regarded as unconventional fodders and have, in recent years, been subjected to chemical analysis for determining their nutritive value. These fodders, however, are common in the mountain areas. Singh (1985) subjected fodder samples taken from nine common fodder trees to chemical composition. He found *Grewia optiva* and *Celtis australis* free of tannins and of high digestibility value. *Grewia optiva* was found to be of exceptionally high nutritive value. He also analysed crop residues obtainable from mountain crops.

Singh *et al.* (1988, 1990) determined nutritive value and digestibility of certain fodder trees occurring in Kumaun Himalaya. These workers established the importance of tree fodder in mountain areas. Singh *et al.* (1995) further underlined the importance of forest-based fodders by reviewing the nutritive value of fodders extracted from a large number of plant species occurring in mountain areas and comparing them with the nutritive values of the conventional fodders. They stressed the need for dependence on uncultivated fodders rather than on cultivated ones not only because of their comparative nutritive values but also for other positive attributes of the system in which such fodder resources are found, *viz.*, higher productivity of fodder per unit area, more resilience, high degree of diversity of fodder-yielding species, sustainability, etc.

Devarajan (1999) reported that the organic matter content of the leaves of ten fodder trees ranged from 76.98 to 95.43 per cent on dry matter basis. Khatta *et al.* (1999) determined the chemical composition of the leaves of various fodder species and reported that the organic matter content varied from 79.0 to 92.0 per cent. Singh (1999) evaluated seasonal variation in dry matter of leaves of five fodder tree species and reported that the dry matter content was highest in winter season (45.0 to 50.0 per cent) followed by rainy (41.5 to 46.0 per cent) and summer season (36.0 to 42.69 per cent). Yadava *et al.* (2003) reported the dry matter and organic matter content of the leaves of 14 plant species to be ranging between 22.0 to 52.70 and 82.52 to 93.52 per cent, respectively.

Crude protein determination of tree fodders has been the sound basis of scientists' claim about the nutritive value of a particular fodder. Sharma *et al.* (1966) determined the crude protein content of *Bauhinia variegata* and *Bambusa arundinacea* leaves and reported that this content was 13.81 and 18.64 per cent on dry matter basis, respectively. Singh (1985) determined the chemical composition of some tree leaves, *viz.*, *Grewia optiva*, *Melia azedarach*, *Bauhinia variegata* and *Toona ciliata* and reported that these species had crude protein content equal to 22.10, 15.35, 15.45 and 14.82 per cent, respectively. Kundu and Sharma (1988) analysed the crude protein content of *Pongamia glabra*, *Ficus religiosa*, *Ficus benghalensis*, *Eugenia jambola*,

Morus indica, Ficus glomerata, Ziziphus nummularia and *Prosopis cineraria* leaves and reported that the crude protein content in these ranged from 8.9 to 19.6 per cent. Devarajan (1999), on the basis of the chemical composition of the leaves of ten fodder tree species, reported crude protein content ranging from 12.88 to 24.38 per cent on dry matter basis. On the basis of his analysis of 20 fodder tree species collected from Eastern Himalaya, Singh (1999) reported that the crude protein content of tree leaves varied from 8.0 to 31.0 per cent on dry matter basis. He also determined the chemical composition of the leaves of 27 bamboo plant species and reported that crude protein content in these species varied from 9.0 to 19.0 per cent on dry matter basis. He also evaluated seasonal variation in nutrient content in the leaves of five fodder species and found that the crude protein content in tree leaves was highest in rainy season (16.94 per cent), followed by summer (15.25 per cent) and winter (13.63 per cent).

Cell wall constituents, *viz.*, neutral detergent fibre (NDF), acid detergent fibre (ADF), hemicelluloses, lignin and cellulose, also determine the nutritive value of fodders. Lohan *et al*. (1980) studied cell wall constituents of the leaves of 26 fodder species and reported that the NDF content was highest in the leaves of *Bambusa nutans* (71.0 per cent) and lowest in the leaves of *Aegle marmelos* (36.4 per cent). They reported that ADF content was highest in the leaves of *Cordia dichotoma* (58.2 per cent) and lowest in the leaves of *Aegle marmelos* (22.5 per cent). The hemicellulose content was highest in the leaves of *Bambusa nutans* (27.0 per cent) and lowest in the leaves of *Cordia dichotoma* (8.8 per cent). The cellulose content was found to be highest in *Robinia pseudoacacia* (37.0 per cent) and lowest in *Leucaena leucocephala* (14.3 per cent). The lignin content was highest in the leaves of *Dendrocalamus hamiltonii* (21.7 per cent) and lowest in the leaves of *Aegle marmelos* (5.2 per cent) on dry matter basis.

There is a long list of animal nutritionists who have analysed cultivated fodders and concentrate feeds for their cell wall constituents, but notable workers amongst those who have attempted the analysis of tree fodders from the Himalayan region in recent years are Bohra (2003, 2006) and Singh and Bohra (2005).

Factors Inducing Range Degradation

Attendant problems facing Himalayan ecosystems are rapid loss of vegetation, the overall quality and performance of socio-economic systems. A survey to the rapid loss of wildlife, including tigers, leopards, hyena–the last *Khedah*, indicating a catastrophic reduction in the number of elephants, in Dehradun district of the Western Himalaya was held in 1905. Excessive querying and its mode of operation and the concentration of cement works have served to turn on of the 'gems' of the Himalaya into a cause celebre of air and noise pollution and serious disruption of water resources and surface cover (Ives and Messerli 1989). Degradation of the rangeland ecosystems is not an isolated phenomenon, but is intrinsically associated with the denudation of fragile mountains with concurrent processes leading to eco-disaster. A spurt of the so-called development programmes in the Himalayan habitats, many of them heavily requiring crumbling and leveling of the mountains for construction activities, have added fuel to fire.

All of these processes have brought in dramatic changes in the Himalayan district of Dehradun over a period of hundred years (1880-1980). A decline in the available arable land per person from about 0.3 ha in 1880 to little over 0.1 ha in 1980. Similarly, per capita access to natural vegetation has declined from about 1.8 ha/ person in 1880 to under 0.4 ha/ person in 1980. There can be no doubt that the district's population has been extensively and progressively impoverished over the hundred-year period, a trend that also implies acceleration in the human pressures on the vegetation of the district (Ives and Messerli 1989).

Declined rangeland productivity as a result of operating factors, triggered processes and accentuating range degradation are critical indicators of range degradation as revealed by several research findings. Over last few decades, ecologically dismal scenarios in the Himalaya are revealed through research findings. For instance, soil erosion rates on sloping lands has increased (20-30 per cent), part of agricultural lands has been abandoned due to decline in land fertility (3-11 per cent), appearance of rocks/ stones on cultivated lands has increased (130-200 per cent), forest area has declined (15-85 per cent), pasture/rangeland area has declined (25-90 per cent), foodgrain production and self-sufficiency has reduced (30-60 per cent) (Shrestha 1992); rate of manure application has decreased (75 per cent), productivity of different crops has gone down (33-75 per cent), livestock holding size has reduced (63-70 per cent), livestock productivity for milk has decreased (50-67 per cent), time of fuel wood, fodder and bedding material has increased (50-80 per cent) (Singh 1992). Shengji (1996), Singh *et al.* (1998), Gaur (2004) and Singh and Gaur (2005) have indicated biodiversity erosion in the Himalayan areas. These persistent negative indicators, indeed, emerge from the ecological degradation of what is the largest component of the mountains' geographical area, *i.e.*, the rangelands.

Improvement of Rangelands

In spite of various constraints to the management of pastures and grasslands, the development of grazing areas and fodder production has tremendous potential in India. Kaul and Ganguli (1963) have recommended that pastures must have 14 per cent of the area under edible bushes to obtain best production results. Silvi-pasture systems on degraded grazing lands have enhanced biomass by up to 7-15 tones/ha/year (Pathak and Roy 1995). Misri (1986) has reported an additional herbage availability of 35-48 tones/ha under horticulture systems.

Singh and Hazra (1995b) have suggested methods for substantial increase in pasture seed production in India. A number of highly productive and disease-resistant and area-specific cultivars of various crop species have been developed by these workers.

Recent years have witnessed some more pragmatic and revolutionary ideas relating to the management of uncultivated lands including rangeland ecosystems. Rangelands are no longer used as a notion of mere grazing lands. "Rangeland management concerns with the multiple uses of rangeland resources rather than seeing rangeland simply as grazing land", argues Zhaoli (2004). At the same time, it recognizes that the public, or more specifically the rangeland users, are important when it comes to making decisions about how rangelands are managed. Rangeland science has shifted from livestock management to rangeland ecology. Rangelands, therefore, have assumed a much wider perspective to be used in their management. While Zhaoli (2004) gives concrete suggestions for the co-management of rangeland resources involving peoples of different HKH countries, Oli (2004) brings to the fore the idea of "corridors" in the management of biodiversity.

Gaur (1999) has enumerated as many as 2150 species belonging to 1032 genera and 189 families in the Garhwal district of Uttarakhand Himalaya. His work underlines the very high degree of biodiversity in mountain habitats. Much of the proportion of this biodiversity is found in the rangeland ecosystems and comprises fodder for livestock. Singh and Gaur (2005) emphasize improvement of rangeland resources for the conservation of rare and endemic biodiversity including wildlife, eco-tourism and for the cohesion amongst peoples of different countries in which the rangelands share the boundaries.

Rangeland ecosystems, in fact, appear to be the most neglected ecosystems of the world despite their maximum coverage on the land area. Nevertheless, in recent years these neglected areas are increasingly capturing the attention of the planners, policy makers and the communities dependent on them. Rangeland resource management is slowly but steadily acquiring adequate attention and intervention. The present study is also an effort towards contributing to rangeland ecology in the parts of the Uttarakhand Himalaya.

Chapter 3
Fodder Diversity and Ethnobotany

The Himalayan rangelands are the natural repositories of biodiversity. The vegetation is predominantly of forest communities with frequent interruption of scrub jungles, savanna, grassy localities and crop fields. Several environmental factors control the distribution of vegetation. However, in the hilly tracts vegetation is demarcated on the basis of altitudinal gradients because edaphic, topographic, climatic and associated factors tend to be altered with altitude. The second important factor in consideration is the aerial distance of the localities from the Greater Himalaya (Gaur 1999).

The Himalayan ranges or peaks are vertical upliftment of mountains of subtropical part. Therefore, instead of commonly used latitudinal terms such as subtropical or tropical and temperate, more appropriately the altitudinal terminology, *viz.*, submontane and montane (Meher-Homji 1973) is used in the following account.

Utilisation of rangelands in the Himalayan region dominated by livestock-dependent communities is primarily for livestock production. These ecosystems harbour a great variety of fodder-providing species. A panorama of the biodiversity presented in this Chapter counts and characterizes the plant species that have some kind of fodder value. A plant species is often of multiple uses. Fodder species are of crucial ethnobotanical values, which have also been mentioned.

The Uttarakhand Himalaya

Uttarakhand (formerly Uttaranchal) was carved out of the northern part of Uttar Pradesh as the 27th state of the Republic of India which came into being on the 9th of November 2000. It is bounded by Uttar Pradesh on the South, Nepal on the East, Himachal Pradesh on the West and Tibet on the Northeast. Geographical area of the state is 53,483 km^2. It extends between 77° 34′ and 81° 02′ E longitude and 28° 43′ to 31° 27′ N latitude. It shares only 1.69 per cent in the area of the country as a whole.

Most of the northern parts of the state are covered by perennial glaciers and alpine pastures while the lower reaches are forested. The topography of Uttarakhand is characterized by hilly terrain, rugged and rocky mountains, deep valleys, high peaks, numerous snow–and rain-fed rivers and rivulets, fragile landmass, limited accessibility, high degree of heterogeneity or diversity, widely scattered habitations, and the likes.

The state has 13 districts, namely Pithoragarh, Champawat, Bageshwar, Almora, Nainital, and Udhamsingh Nagar in the Kumaun Division and Chamoli, Rudraprayag, Uttarkashi, Pauri, Tehri, Dehradun and Haridwar in the Garhwal Division. There are 15761 inhabited villages in Uttarakhand.

Most of the land in Uttarakhand is under forests (62.6 per cent). Net sown area is only 14.3 per cent and 3.8 per cent area is under miscellaneous tree crops and groves, while the cultivable wasteland is 5.7 percent, which is too high (Teli 2006). A general picture of land use in the Uttarakhand state is depicted in Figure 3.1.

According to the census of India 2001, Uttarakhand has 8.48 million population with a density of 159 persons per km^2. Out of this population, 4.32 million are men and 4.16 million women. Rural population comprises 74.41 per cent of the total population. The decennial population growth rate in Uttarakhand over 1991-2001 was recorded to be 19.20 per cent. Literacy rate is 84.01 and 60.26

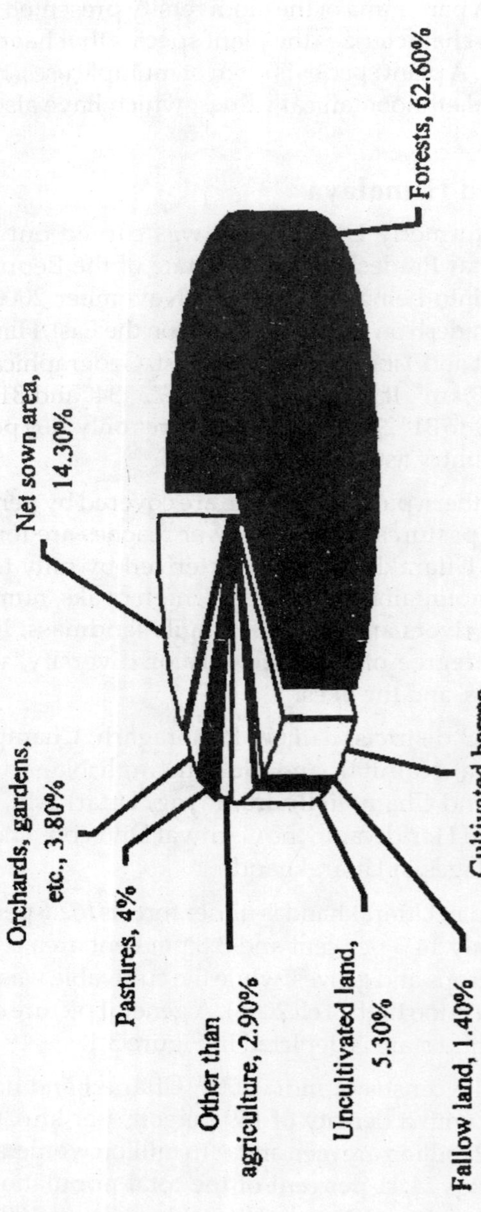

Figure 3.1: Land Use in Uttarakhand State

per cent amongst male and female populations, respectively. About 85 per cent of the working population is engaged in primary activities.

Cultural heterogeneity (which is regarded to be the manifestation of natural diversity) is also one of the unique features of Uttarakhand. Following chiefly the agrarian-pastoral way of life the mountain communities embrace many colours of ethnicity. A sizeable majority occupying areas in the vicinity of Greater Himalaya comprises tribal people. Bhotiyas, Marchhas, Tolchas, Koltas, Jaunsaris, Van Rauts, etc. are the notable tribes. Gaddis (coming from Himachal Pradesh) and Gujjars (residing in the foothills of Uttarakhand during winter season) are the nomads. Most of the tribes follow transhumance. Nomads and transhumant tribes own large number of livestock, both bovine and ovine species, and depend on rangelands spread over vast stretches of the mountains for their livelihoods. The rangelands in the Uttarakhand state support some 2188 thousand cattle, 1228 thousand buffaloes, 296 thousand sheep, 1158 thousand goats, 17 thousand horses and ponies and some 22 thousand mules. This huge population of livestock is a lifeline for the majority of mountain population.

Field Survey, Plant Collection, Preservation and Maintenance

Enumeration of biodiversity in the rangelands presented in this Chapter is based on the extensive survey and collection of range plants particularly from Almora and Champawat districts of Kumaun Himalaya and Pauri of Garhwal Himalaya. For the collection of plants and preparation of taxonomic accounts, frequent field trips were made during different seasons of three consecutive years (2003-2006). Several remote localities, ranging from sub-montane to montane were covered for investigation. The plants were pressed inside the blotters or newspapers and dried by successive changes. Detailed information regarding locality, season, distribution, vernacular names, availability and general association were recorded in each case in the field.

The well pressed and dried specimens were poisoned with $HgCl_2$ solution, prepared in saturated ethyl alcohol or denatured spirit. After poisoning, the dried specimens were mounted on the standard (42x28cm) Herbarium sheets, following the techniques as suggested by Jain and Rao (1977).

The plant specimens were identified with the help of recent and relevant floras and revisions, and compared with the authentic Herbarium specimens of Botanical Survey of India, Northern Circle Dehradun (BSD), and HNB Garhwal University Herbarium (GUH), Srinagar for their information. Each specimen was labelled with botanical name, date or month of collection, vernaculars, habit and habitat, distribution, approximate elevation, collector's name and other necessary remarks.

Diversity of Fodder Species

The rangelands which were covered for this study were located in sub-montane and montane areas. About 550 species of fodder-yielding plants including grasses, other herbaceous plants, trees and shrubs were found in the submontane and montane rangelands in the mountains of Uttarakhand (Table 3.1). This variety of fodders available in the rangelands is amazing particularly when no more than a dozen fodder species are conventionally referred to and recommended for mountain areas (Singh *et al.*, 1995).

The range vegetation determines the ecosystem quality and has phenomenal influence on the performance of agriculture. Animal husbandry, which is the chief occupation of the mountain communities, is directly dependent on the rangelands. The local communities, therefore, attempt to manage the vegetation which has direct bearing on animal husbandry and other livelihood activities. Woody perennials as well as herbaceous plants contribute to the livestock fodder supplies. Some plants are grazed or browsed only by ovine species or consumed only by non-ruminants, like rabbits, not by all livestock species. Leaves and small tender twigs of certain trees and shrubs are lopped for feeding livestock at stall.

Singh *et al.* (1995) and Singh and Bohra (2005) have given the list of 50 fodder trees, shrubs and non-graminaceous herbaceous plants occurring in mountain habitats. These fodder species occur in the agro-forestry systems or in the tree-dominated rangelands. Bohra (2006) has enumerated some 160 range plants that provide fodder in parts of the Kumaun Himalaya.

As is clearly revealed from the list of fodder plants presented in Table 3.1, most of the plants constitute grasses belonging to the family Poaceae. The grass-dominated rangelands especially harbour a

Table 3.1: Diversity of the Fodder Species in the Rangeland Ecosystems and their Ethnobotanical Values

Botanical Name	*Vernacular Name (English Name)*	*Family*	*Site/ Location*	*Altitude, m msl*	*Ethnobotanical Notes*
Acacia catechu (L.f.) Willd.	*Khair, Katha* (Cutch Tree)	Mimosaceae	Kosi, Almora	1250	Hard and durable wood variously used, particularly as a charcoal by blacksmiths. Wood yields 'Katha' or 'Catechu' of commercial value, used for various medicines, particularly in digestive and respiratory diseases; bark used in diarrhoea, dysentery, bronchitis and menstrual disorders; leave make fodder for goats.
Acacia farnesiana (L.) Willd.	*Vilayati Kikar* (Cassie Flower, Opopanax)	Mimosaceae	Byasghat,	Pauri	1000 Bark regarded as an astringent and effective medicine for gonorrhoea. Plant extract also used against snake bite, rabies and delirium. It yields a gum similar to gum-arabic. Flowers often used in perfumes. Plant is a good soil-binder. Leaves supposed to increase quantity and quality of milk in livestock.
Achyranthes bidentata Blume	*Chicheree, Latjira*	Amaranthaceae	Abattmount, Lohaghat	1700	Root infusion taken in malarial fever; leaf extract supposed to facilitate delivery; root powder used in making local beverage; plant decoction used in dropsy and bronchitis.
Aconogonum molle (D. Don) Hara	*Ameta*	Polygonaceae	Kosi, Almora	1300	Plant used as fodder.
Adhatoda vasica Nees	*Basing, Baisingu*	Acanthaceae	Abattmount, Lohaghat	1700	Young twigs made into vegetable, supposed to be good for cough and cold; juice of flowers useful in pulmonary affections; leaves and roots infusion useful in bronchitis and fever; flowers as a source of bee-forage.

Contd...

Table 3.1–Contd...

Botanical Name	*Vernacular Name (English Name)*	*Family*	*Site/ Location*	*Altitude, m msl*	*Ethnobotanical Notes*
Adina cordifolia (Roxb.) Hook. f. ex Brandis	*Haldu* (Yellow Teak)	Rubiaceae	Almora	1500	Wood yellowish, used for construction, furniture, agricultural implements, combs and several articles; leaves as fodder; decoction of bark with whey taken in diabetes; roots chewed in dysentery; an important tree of social forestry.
Aechmanthera gossypina (Nees) Nees	*Jaundela, Bairara*	Acanthaceae	Bhanargaon	1400	Bark fibre used for ropes and netting; flowers as source of bee-forage.
Aesculus indica (Colebr. ex Cambess.) Hook.	*Pangar* (Himalayan Horse Chestnut)	Hippocas-tanaceae	Pauri, Khirsu	2200	Cream coloured wood made into pots and vessels ; fruits given to cattle and goats; flour from seeds mixed with wheat flour during famine; seed paste and oil applied in rheumatic pain; flowers useful in apiculture as bee-forage; leaves as fodder and manure; an important tree of social forestry.
Agrimonia pilosa Ledebour	*Lesukuria, Velu*	Rosaceae	Patan, Lohaghat	1650	Plant decoction given in cough and diarrhoea; root paste with mustard oil applied round the belly in suppressed urination.
Agropyron longe-aristatum Boiss.		Poaceae	Abattmount, Lohaghat	1700	Grazed by animals.
Agrostis nervosa Nees ex Trinius		Poaceae	Chaubatiya	2300	Fodder for livestock.

Contd...

Table 3.1–Contd...

Botanical Name	*Vernacular Name (English Name)*	*Family*	*Site/ Location*	*Altitude, m msl*	*Ethnobotanical Notes*
Agrostis pilosula Trinius		Poaceae	Chauri, Lohaghat	1650	Commonly grazed by livestock.
Agrostis vinealis Schreber		Poaceae	Kosi, Almora	1250	Fodder for livestock.
Albizia chinensis (Osbeck) Merrill	*Kala-siris, Siran*	Mimosaceae	Kotdwar, Rishikesh	600	Foliage lopped for fodder; wood used for various purposes, especially for packaging cases. Flowers as source of bee-forage.
Albizia julibrissin Durazzini	*Sirsa, Kaunera, Bukarna* (Pink Siris)	Mimosaceae	Almora	1600	Foliage lopped for fodder; wood used for various purposes; flowers as source of bee-forage; an important tree of social forestry.
Albizia lebbeck (L.) Benth.	*Siris, Siras* (Siris Tree)	Mimosaceae	Mahatgaon, Almora	1400	Leaves provide palatable fodder to cattle; wood used for various articles and furniture; seeds and bark used in diarrhea, dysentery, eye complaints and in piles; gum exuded used in ulcer; flowers as source of bee-forage; an important tree of social forestry.
Albizia lucidior (Steudel) Nielsen		Mimosaceae	Cheela	800	Occasional fodder from leaves; wood for various purposes.
Albizia odoratissima (L.f.) Benth.	*Bansa, Bassein, kala Siris*	Mimosaceae	Corbette Park	800	Foliage lopped for fodder; wood used for making several articles and furniture. Bark occasionally used for poisoning fishes; important tree of social forestry.

Contd...

Table 3.1–Contd...

Botanical Name	*Vernacular Name (English Name)*	*Family*	*Site/ Location*	*Altitude, m msl*	*Ethnobotanical Notes*
Albizia procera (Roxb.) Benth.	*Karah, Karahi, Safed Siris*	Mimosaceae	Laldhang	600	Wood used for various purposes; bark used for tanning and to intoxicate fishes; seed powder used in gonorrhoea; flowers as source of bee-forage; occasionally used for fodder; an important tree of social forestry.
Alloteropsis cimicina (L.) Stapf		Poaceae	Bhanargaon, Almora	1400	As fodder.
Alternanthera sessilis L.	*Gudre-Saag*	Amaranthaceae	Kosi, Almora	1250	Occasionally leaves used as vegetable during scarcity.
Alysicarpus bupleurifolius (L.) DC.		Fabaceae	Srinagar	1000	Fodder for cattle.
Alysicarpus heyneanus Wight and Arn.		Fabaceae	Srinagar	1000	Fodder for cattle.
Alysicarpus procumbens (Roxb.) Schindler		Fabaceae	Kotdwar	400	Occasionally used as fodder.
Alysicarpus vaginalis (L.) DC.		Fabaceae	Srinagar	1000	Occasionally used as fodder.
Amaranthus spinosus L.	*Kantelu-Marsu* (Prickly Amaranth)	Amarantha-ceae	Mahatgaon, Almora	1400	Leaves used for vegetable; infusion of leaves with salt is given in dysmenorrhoea; plant also used as fodder.

Contd...

Table 3.1–Contd...

Botanical Name	Vernacular Name (English Name)	Family	Site/ Location	Altitude, m msl	Ethnobotanical Notes
Amaranthus viridis L.	*Jangli Chulai* (Wild Amaranth)	Amarantha-ceae	Kosi, Almora	1300	Young shoots and leaves cooked as vegetable.
Ampelocissus divaricata (Wallich ex Lawson) Planchon	*Pan Laguli*	Vitaceae	Katarmal, Almora	1650	Fruits and young leaves edible; stem as fibre.
Ampelocissus rugosa (Wallich) Planchon	*Chhipari, Bhel-Chapru*	Vitaceae	Chaubatiya, Almora	2000	Paste of leaves and stem applied in bone fracture; ripe fruits edibie.
Anaphalis cinnamomea C.B. Clarke	*Bugla, Bugliya*	Asteraceae	Abattmount, Lohaghat	1700	Paste of heads and leaves applied on cuts, wounds and boils; fibres of leaves and stem used to lit the fire by friction of stones, locally known as 'Agela.'
Andropogon munroi C.B. Clarke	*Musliya-ghas*	Poaceae	Daira-Binsar	2400	Used as fodder.
Anemone vitifolia Buch.-Ham. ex DC.	*Mudeela* (Vine Leaved Anemone)	Ranuncula-ceae	Khirsu, Binsar	2400	Roots' and leaves' paste applied externally on ringworm and eczema in folk medicine; leaves used as fodder.
Anogeissus latifolius (Roxb. ex DC.) Wallich ex Richard	*Dhauri, Dhawadi*	Combreta-ceae	Bhanargaon	1400	Wood used for construction and agricultural implements ; gum of the tree useful as a source of tannin ; leaves yield black dye; flowers as a source of bee-forage in apiculture; sometimes leaves act as host of tasar silkworm; useful plant of social forestry.

Contd...

Table 3.1–Contd...

Botanical Name	*Vernacular Name (English Name)*	*Family*	*Site/ Location*	*Altitude, m msl*	*Ethnobotanical Notes*
Apluda aristata L.		Poaceae	Sinagar	1200	As fodder at young stage.
Apluda mutica L.	*Tachula, Ponai, Bhanjura*	Poaceae	Chanmari, Lohaghat	1650	For thatching huts and as brooms.
Argyrolobium roseum (Cambess.) Jaubert and Spach		Fabaceae	Matyali	1600	Plant browsed by cattle.
Aristida adscensionis L.		Poaceae	Srinagar	1200	Grazed by cattle.
Artemisia nilagirica C.B. Clarke	*Kunjaa*	Asteraceae	Abattmount, Lohaghat	1700	Leaf juice used against intestinal worms; young twigs made into brooms; plant regarded as sacred.
Artemisia parviflora Roxb.	*Patee, Pamasi*	Asteraceae	Abattmount, Lohaghat	1700	Leaves and flower tops mixed with butter and used as incense, as well as an insecticide; plant regarded as sacred.
Arthraxon hispidus (Thumb.) Makino		Poaceae	Binsar	2200	Browsed by animals.
Arthraxon lanceolatus (Roxb.) Hochst.		Poaceae	Katarmal, Almora	1550	Used for fodder.

Contd...

Table 3.1–Contd...

Botanical Name	Vernacular Name (English Name)	Family	Site/ Location	Altitude, m msl	Ethnobotanical Notes
Arthraxon lancifolius (Trinius) Hochst.		Poaceae	Pauri	1600	Grazed by animals.
Arthraxon lancifolius (Trinus) Hochst.	*Kangulya*	Poaceae	Chauri, Lohaghat	1650	Fodder for cattle.
Arthraxon prionodes (Steudel) Dandy		Poaceae	Khirsu	2300	Browsed by animals.
Arthraxon prionodes (Steudel) Dandy	*Ghas, Ghaa*	Poaceae	Abattmount, Lohaghat	1700	Fodder for cattle.
Arundinella bengalensis (Sprengel) Druce		Poaceae	Almora	1500	Fodder for livestock.
Arundinella birmanica Hook. f.		Poaceae	Pauri	1600	Grazed by animals.
Arundinella nepalensis Trinius		Poaceae	Chanmari Lohaghat	1650	Used for thatching and for fodder.

Contd...

Table 3.1–Contd...

Botanical Name	*Vernacular Name (English Name)*	*Family*	*Site/ Location*	*Altitude, m msl*	*Ethnobotanical Notes*
Arundinella nervosa (Roxb.) Nees ex Hook. and Arn.		Poaceae	Khirsu	2300	Grazed by animals.
Arundinella pumila (Hochst. ex Richard) Studel		Poaceae	Pauri	1600	Grazed by animals.
Arundo donax L.	*Tinta, Nar* (Great Reed)	Poaceae	Khirsu	2300	Eaten by cattle; brooms are prepared from the panicles ; culms used in thatching and making baskets.
Aster molliusculus Lindley ex DC.) C.B. Clarke	*Phulyan, Phulari*	Asteraceae	Khetikhan	1650	Root and leaf powder used in gastric problems.
Astilbe rivularis Buch.-Ham. ex D.Don	*Pothee, Gosy*	Saxifraga-ceae	Ghingharikhal, Almora	1950	Root powder believed to be as an abortifacient; leaves as fodder.
Astragalus chlorostachys Lindley	*Rudravanti*	Fabaceae	Kodiabagarh	3000	Pods edible; leaves as fodder. Plant infusion taken as tonic as well as febrifuge; decoction of root given in tuberculosis.

Contd...

Table 3.1–Contd...

Botanical Name	Vernacular Name (English Name)	Family	Site/ Location	Altitude, m msl	Ethnobotanical Notes
Astragalus graveolens Buch.-Ham. ex Benth.		Fabaceae	Adhwani	2400	Leaves used as fodder.
Astragalus leucocephalus Graham ex Benth.	*Rudravanti*	Fabaceae	Pauri	2200	Leaves occasionally used as fodder.
Avena fatua L.	*Javatu, Jai*	Poaceae	Almora	1650	Fodder for livestock.
Bambusa arundinacea Willd.	*Kantna-Bans* Thorny Bamboo	Poaceae Champawat	Tanakpur,	1200	Multipurpose bamboo; stem used for thatching roofing, construction, furniture, etc.; leaves as fodder; young twigs pickled or used as vegetable.
Barleria cristata L.	*Saundi, Kala Bansa*	Acanthaceae	Katarmal, Almora	1600	Root decoction used against bronchitis and pneumonia; leaves and root paste applied on wound swelling; root chips added to local beverages; seeds regarded as antidote to snake bite; flowers as source of bee-forage; plants act as soil binder.
Basella rubra L.	*Lal-bachlu, Poi*	Basellaceae	Abattmount, Lohaghat	1700	Chiefly used as pot herb.
Bauhinia acuminata L.	*Kachnar*	Caesalpiniaceae	Kosi, Almora	1250	Leaves used as fodder; branches yield fibre and wood as fuel and timber.

Contd...

Table 3.1–Contd...

Botanical Name	*Vernacular Name (English Name)*	*Family*	*Site/ Location*	*Altitude, m msl*	*Ethnobotanical Notes*
Bauhinia purpurea L.	*Guiral, Kwiryal* (Purple Bauhinia)	Caesalpinia-ceae	Kosi, Almora	1250	Flower buds cooked as vegetable and foliage used for agricultural implements; bark used in dyeing and tanning, as well as in medicine as detergent of wounds; important tree of social forestry.
Bauhinia racemosa Lam.	*Jhinghora*	Caesalpinia-ceae	Mahatgaon	1400	Leaves used as fodder; bark, leaves and fruits medicinal; bark used as fibre; an important tree of social forestry.
Bauhinia semla Wunderlin	*Semla, Samlu, Kundlai, Kanda*	Caeselpinia-ceae	Dugadda	800	Leaves used as fodder; wood as timber and fuel. The gum of the plant resembles gum-arabic and sold in the market, used medicinally; important tree of social forestry.
Bauhinia vahlii Wight and Arn.	*Malu*	Caesalpinia-ceae	Tarag, Lohaghat	1650	Leaves made into plates, cups, umbrellas and grain storage-sacs; fibre obtained from the bark variously used, especially to bind mats and temporary portable huts (*tat and pal*); wood as fuel; leaves lopped for fodder.
Bauhinia variegata L.	*Kwaral, Guriyal*	Caesalpinia-ceae	Matela, Almora	1200	Leaves provide good quality fodder; fibre from bark variously used; young flowers eaten as vegetable; wood used for agricultural implements and construction; ash of dried leaves taken in cough; an important tree of social forestry.
Berberis aristata DC.	*Kingore, Kasmor* (Indian Barberry)	Berberida-ceae	Chaubatiya, Almora	2250	Juice from bark of stem or root often known as 'Rasaut' dropped in ophthalmia; infusion of root given in fever; fruits edible; bark yields dye; leaves rarely browsed by goats.

Contd...

Table 3.1–Contd...

Botanical Name	*Vernacular Name (English Name)*	*Family*	*Site/ Location*	*Altitude, m msl*	*Ethnobotanical Notes*
Berberis asiatica Roxb. ex DC.	*Kilmora, Kingor* (Asiatic Barberry)	Berberidaceae	Katarmal, Almora	1650	Root extracts in ophthalmia; fruits edible; root and bark yield yellow dye; fruits edible; leaves can be browsed by goats.
Berchemia floribunda (Wallich) Brongniart	*Amili, Kouloi*	Rhamnaceae	Sara-Yamkeshwar	1200	Mature leaves used as fodder; root decoction in fever; flowers as source of bee-forage.
Betula alnoides Buch.-Ham. ex D. Don	*Saur, Kath-bhuj*	Betulaceae	Abattmount, Lohaghat	2200	Wood used for furniture and cabinet work; bark with hot butter used as snake antidote. Leaves lopped for fodder.
Bidens biternata (Lour.) Merril and Sherff	*Mangrinya*	Asteraceae	Kosi, Almora	1250	Vegetable of leaves supposed to be good at the initial stages of leprosy and keeps away from several ailments if women folk take the vegetable and infants use breast feeding; leaf juice used in cuts.
Bidens pilosa L.	*Kumra, Kurei*	Asteraceae	Khetikhan, Lohaghat	1650	Plant extract with honey used in cough and bronchitis; vegetable of the herb considered useful in leucoderma.
Boehmeria macrophylla D.Don	*Bara-Siauru*	Urticaceae	Kosi, Almora	1200	Leaves provide fodder; bark yields strong fibre.
Boehmeria platyphylla D.Don	*Khagsa*	Urticaceae	Chanmari, Lohaghat	1650	Plant regarded as a good fodder to cattle; bark yields strong cordage fibre.

Contd...

Table 3.1–Contd...

Botanical Name	*Vernacular Name (English Name)*	*Family*	*Site/ Location*	*Altitude, m msl*	*Ethnobotanical Notes*
Boehmeria rugulosa Wedd.	*Genthi*	Urticaceae	Katarmal, Almora	1700	Wood often used for vessels, combs, boxes, cups and other articles; leaves provide good palatable fodder; plaster of bark applied on fractured bones; important tree of social forestry.
Boenninghausenia albiflora (Hook.) Reichb. ex Meisn.	*Pishumar, Upniyaghas*	Rutaceae	Abattmount, Lohaghat	1700	Leaf paste applied on cuts and wounds; root powder used as antiseptic and juice given in vomiting and dysentery.
Boswellia serrata Roxb. ex Colebr.	*Salar, Salai* (Olibanum Tree)	Burseraceae	Kosi, Almora	1250	Transparent gum-resin exuded from the bark burnt as incense and also used in skin and digestive disorders.
Bothriochloa foulkesii (Hook.f.) Henrard		Poaceae	Rewakhet, Pauri	1400	Fodder for livestock.
Bothriochloa glabra (Roxb.) *A. Camus*		Poaceae	Binsar, Pauri	1800	Fodder for livestock.
Bothriochloa ischaemum (L.) Keng		Poaceae	Srinagar, Pauri	1200	Fodder for livestock.
Brachiaria distachya (L.) Stapf		Poaceae	Srinagar	1000	Fodder for cattle.
Brachiaria ramosa (L.) Stapf		Poaceae	Matela, Almora	1200	Fodder for livestock.

Contd...

Table 3.1–Contd...

Botanical Name	*Vernacular Name (English Name)*	*Family*	*Site/ Location*	*Altitude, m msl*	*Ethnobotanical Notes*
Brachiaria villosa (Lam.) *A. Camus*	*Malchu*	Poaceae	Pauri	1600	Used as fodder for livestock.
Brachypodium sylvaticum (Hudson) *P. Beauv.*		Poaceae	Chauri, Lohaghat	1650	Used for fodder.
Bridelia retusa (L.) Sprengel	*Gauli, Ekdaniya*	Euphorbiaceae	Laldhang	1000	Bark yields tannin; leaves as fodder; fruits edible, sweet in taste; decoction of roots given in diarrhoea; bark in boiled water is used to gargle in gum diseases.
Bridelia verrucosa Haines	*Goya, Gondni, Kangnalia*	Euphorbiaceae	Dugadda	1000	Leaves lopped for fodder; roots believed to act as vermifuge.
Bromus catharticus Vahl	*Brome*	Poaceae	Pauri	1600	Used as fodder for cattle.
Buchanania latifolia Roxb.	*Maira, Piyal*	Anacardiaceae	Mahatgaon, Almora	1450	Wood used for making boxes, doors and household furniture; fruits edible; leaves used as substitute of plates; gum exuded from stem used in indigenous medicines; paste of kernel good for cough and cold and also used in skin ailments.
Bupleurum hamiltonii Balakrishnan	*Jangli-jeera*	Apiaceae	Abattmount, Lohaghat	1700	Roots used in stomach and liver disorders.

Contd...

Table 3.1–Contd...

Botanical Name	*Vernacular Name (English Name)*	*Family*	*Site/ Location*	*Altitude, m msl*	*Ethnobotanical Notes*
Butea monosperma (Lam.) Kuntze	*Dhak, Tesu, Palas* (Flame of the Forest)	Fabaceae	Kotdwar	400	Leaves as fodder; also used for making cups and plates. Flowers yield yellow dye, used variously. Decoction of flowers regarded as blood purifier. Seeds as antiseptic and anthelmintic. Gum exudes used as tonic and in indigestion; timber for construction and for making into several articles; an important tree of social forestry.
Caesalpinia bonduc (L.) Roxb.	*Karaunj, Kanja* (Fever Nut)	Caesalpiniaceae	Matela, Almora	1200	Seeds used in colic, intermittent fever and in disturbed menstruation; cotyledons as substitute of Quinine; regarded good fodder for goats and sheep.
Caesalpinia decapetala (Roth) Alston	*Kingari, Kunju* (Bahama Sappan)	Caesalpiniaceae	Mahatgaon, Almora	1400	Leaves used as fodder of goats and sheep; leaves applied to burns; bark yields tannin; flowers useful source of bee-forage.
Capillipedium assimile (Steudel) *A. Camus*		Poaceae	Srinagar	1200	Culms sometimes used by pipe smokers; plant is a good fodder.
Capillipedium assimile Steudel		Poaceae	Abattmount, Lohaghat	1700 fodder.	Culms sometimes used by pipe smokers; plant is a good
Capillipedium huegelii (Hackel) Stapf		Poaceae	Gumkhal	1600	As fodder for livestock.
Capillipedium parviflorum (R.Br.) Stapf	*Nigaliaghas*	Poaceae	Chauri, Lohaghat	1650	Fodder for sheep, cattle and goats.

Contd...

Table 3.1–Contd...

Botanical Name	Vernacular Name (English Name)	Family	Site/ Location	Altitude, m msl	Ethnobotanical Notes
Capillipedium parviflorum (R.Br.) Stapf	*Nigaliaghas*	Poaceae	Binsar-Palan	2400	Fodder for sheep, cattle and goats.
Capparis sepiaria L.	*Kanthari, Karva* (Capu)	Capparaceae	Matela, Almora	1200	Leaves and twigs browsed by sheep and goats; fruits edible; decoction of roots used in paralytic attacks.
Capparis zeylanica L.	*Bauri*	Capparaceae	Dhaman	800	Leaves and young twigs browsed by sheep and goats; root bark medicinal, especially in diabetes; flowers as a source of bee-forage in apiary.
Carex cruciata Wahlenberg		Cyperaceae	Chanmari Lohaghat	1650	Occasionally as fodder.
Carpinus viminea Lindley	*Putli, Chamkharik*	Corylaceae	Chaubatiya	2300	Wood used for furniture and articles of sports as well as weaving shuttle; leaves used as fodder.
Caryopteris foetida (D. Don) Thellung	*Karwi*	Verbenaceae	Adhwani, Binsar	2300	Used as fodder; leaf juice applied on wounds; flowers as source of bee-forage.
Caryopteris odorata (D. Don) B.L. Robinson		Verbenaceae	Kaleshwar, Srinagar	1200	Twigs and leaves used as fodder; decoction of plant considered as an abortifacient.
Cassia absus L.	*Chaaksee, Cheaksu, Chaksi*	Caesalpiniaceae	Sara, Srinagar	1000	Seeds used in ophthalmic ailments and skin diseases; leaf juice given in diarrhoea.

Contd...

Table 3.1–Contd...

Botanical Name	*Vernacular Name (English Name)*	*Family*	*Site/ Location*	*Altitude, m msl*	*Ethnobotanical Notes*
Cassia mimosoides L.		Caesalpiniaceae	Chanmari, Lohaghat	1650	Leaves in skin diseases; fruits believed to be an antidote of snake bite; decoction of root given in dropsy; leaf and root paste useful in piles, boils and ringworms.
Cassia tora L.	*Chakunda, Panvar, Chakramarde*	Caesalpiniaceae	Patan, Lohaghat	1650	Seeds and leaves used in skin diseases, cuts and wounds and in bone fracture; leaves and roots also used as vermicide and febrifuge; leaves increase quantity and quality of milk in livestock.
Castanopsis tribuloides (J.E. Smith) A. DC.	*Katonj*	Fagaceae	Lansdowne, Pauri	2000	Wood used for construction; fruits edible.
Catunaregam uliginosa (Retz.) Sivarajan	*Gangati*	Rubiaceae	Kotdwar	400	Leaves as fodder; seed paste used in skin ailments; flowers as source of bee-forage.
Cayratia trifolia (L.) Domin	*Amal-bel* (Fox Grape)	Vitaceae	Kosi, Almora	1250	Leaves as fodder; poultice of leaves used for the treatment of boils; plant used in many ways in Ayurvedic system.
Celtis australis L.	*Kharik*	Urticaceae	Tarag, Lohaghat	1650	Fruits edible; leaves provide good fodder; bark gives yellow dye; wood used for making small articles; bark paste applied on bones, pimples, contusions, sprains and joint pains; an important agroforestry tree.
Cenchrus ciliaris L.	*Anjan, Dhaman*	Poaceae	Kotdwar	400	As fodder.

Contd...

Table 3.1–Contd...

Botanical Name	*Vernacular Name (English Name)*	*Family*	*Site/ Location*	*Altitude, m msl*	*Ethnobotanical Notes*
Chenopodium album L.	*Bathua, Bathu* (Lamb's Quarters)	Chenopodiaceae	Srinagar, Pauri	1400	Leaves used as vegetable; breads prepared from the ground grains; contributes to increase quantity and quality of milk.
Chenopodium ambrosioides L.	(Mexican Tea)	Chenopodiaceae	Chauri, Lohaghat	1550	Leaves used as vegetables; contributes to increase quantity and quality of milk.
Chenopodium botrys L.	Kirjauri-jhar (Sticky Goosefoot)	Chenopodiaceae	Thailisain, Pauri	2200	Leaves used as vegetables.
Chenopodium foliosum (Moench) Ascherson	Bathu (Goosefoot)	Chenopodiaceae	Binsar, Pauri	2300	Leaves used as vegetables.
Chenopodium murale L.	(Nettle Leaved Goose-foot)	Chenopodiaceae	Srinagar, Pauri	1000	Leaves used as vegetables; contributes to increase quantity and quality of milk.
Chloris dolichostachya Lagasca		Poaceae	Rishikesh, Srinagar	600	As fodder.
Chrysanthemum leucantheum L.		Asteraceae	Chanmari, Lohaghat	1650	Occasionally as fodder.
Chrysopogon aciculatus (Retz.) Trinius	*Surwala, Lampa*	Poaceae	Srinagar	800	As fodder but sharp callus intermingled with the body of cattle.

Contd...

Table 3.1–Contd...

Botanical Name	*Vernacular Name (English Name)*	*Family*	*Site/ Location*	*Altitude, m msl*	*Ethnobotanical Notes*
Chrysopogon fulvus (Sprengel) Chiovenda	*Gurla, Goria, Zargha*	Poaceae	Srinagar	1000	As fodder.
Chrysopogon gryllus (L.) Trinius	*Salum, Salima*	Poaceae	Chauri. Lohaghat	1650	Contributes to increase milk quantity and quality.
Cinnamomum tamala (Buch.-Ham.) Nees and Ebermaeir	*Dalchini, Tejpat*	Lauraceae	Abattmount, Lohaghat	1700	Bark and dried leaves used for flavouring tea and various food preparations; bark medicinal, chewed in dyspepsia and throat irritation.
Cissampelos pareira L.	*Pahre, Parha* (False Pareira)	Menisperma-ceae	Kotdwara, Srinagar, Satpuli	1200	Leaves and roots believed as an antidote to snake or scorpion bite; root extract used in constipation and gastric troubles; flowering twigs in psychotherapy; roots also used in cough and urinary troubles; helps in increasing milk yields and quality.
Cissus repanda Vahl	*Bhel-Chapru, Pani-bel*	Vitaceae	Thailsain	1200	Sap of stem used to wash sores and wounds; fruits edible; leaves as fodder.
Clematis barbellata Edgew	*Kanrya, Santai*	Ranuncula-ceae	Kodiyabagarh, Pauri	2600	Stem used for coarse ropes; decoction of leaves applied on scabies; used to increase quantity and quality of milk in livestock.
Clematis bucha-naniana DC.	*Kanguli, Lagulia*	Ranuncula-ceae	Chauri, Lohaghat	1650	Leaf paste externally applied for skin ailments; leaves sometimes supposed to be poisonous for cattle.

Contd...

Table 3.1–Contd...

Botanical Name	*Vernacular Name (English Name)*	*Family*	*Site/ Location*	*Altitude, m msl*	*Ethnobotanical Notes*
Clematis connata DC.		Ranuncula-ceae	Binsar	2500	Leaves as fodder to goats and sheep.
Clematis grata Wallich	*Laguli, Santai, Dhanwali*	Ranuncula-ceae	Neelkanth, Bharsar	2700	Leaves as fodder especially for goats and sheep; used for increasing quantity and quality of milk.
Codariocalyx motorius (Houttuyn) Ohashi		Fabaceae	Srinagar	1200	Leaves lopped for fodder; plant is a good soil-binder.
Coix lacryma-jobi L.	*Sankuru* (Adlay, Job's Tears)	Poaceae	Thailisain	1000	Leaves as fodder; grains for beads and rosaries, rarely as substitute of rice.
Commelina benghalensis L.	*Kanjula, Kansura*	Commelina-ceae	Abattmount, Lohaghat	1700	Plant juice given in dysentery and paste applied on body swelling and ache.
Coriaria nepalensis Wallich	*Makroli, Masuri* (Mussoorie Berry)	Coriaria-ceae	Khirsu, Chaurikhal, Pauri	2300	Fruits edible, also given as an emetic; leaves and bark used to intoxicate fishes; baskets are made from the stem.
Cornus oblonga Wallich	*Katkanai*	Cornaceae	Abattmount, Lohaghat	1700	Wood useful for agricultural implements; leaves provide fodder.
Cotoneaster lindleyi Steudel	*Dhuis*	Rosaceae	Kundinagar, Binsar	2600	Ripe fruits edible; wood as fuel; leaves as fodder.

Contd...

Table 3.1–Contd...

Botanical Name	*Vernacular Name (English Name)*	*Family*	*Site/ Location*	*Altitude, m msl*	*Ethnobotanical Notes*
Crataeva adansonii DC.	*Barne*	Capparaceae	Kotdwar	500	Leaves as fodder; wood used for agricultural implements, drums, models and combs. Fruits edible, often pickled. Leaves and bark medicinal in treatment of bronchitis, urinary and skin disorders; flowers as source of bee-forage.
Crotalaria medicaginea Lam.	*Ban-Methi, Gulabi*	Fabaceae	Mahatgaon	1400	Leaf juice applied in scabies and urticaria; contributes to increase milk yields and quality.
Cuscuta reflexa Roxb.	*Akash-laguli*	Convolvulaceae	Abattmount, Lohaghat	1700	Poultice of stem used in rheumatic pain and swellings and also other skin ailments; flowers source of bee-forage.
Cyananthus lobatus Wallich ex Benth		Campanulaceae	Kodiabagarh, Pauri	3000	Increases milk quantity and quality in livestock.
Cyathula tomentosa (Roth) Moq.	*Lichkura, Silla*	Amaranthaceae	Almora	1600	Leaf extract has emetic properties and given in snake bite; plant used as fodder.
Cymbopogon flexuosus (Nees ex Steudel) Watson		Poaceae	Dhikala	600	As fodder for livestock and wild animals.
Cymbopogon jwarancusa (Jones) Schultes	*Tintyra*	Poaceae	Srinagar, Sara	1000	Roots are chewed in soar throat; occasionally grazed by animals.

Contd...

Table 3.1–Contd...

Botanical Name	Vernacular Name (English Name)	Family	Site/ Location	Altitude, m msl	Ethnobotanical Notes
Cymbopogon martinii (Roxb.) W. Watson	*Piriya-ghas* (Rusa Grass, Rohsa Grass)	Poaceae	Srinagar, Chipalghat	1100	Crushed leaves smelled in cold; leaves are burnt as incense; occasionally browsed by animals.
Cynodon dactylon (L.) Persoon	*Dubla, Doob*	Poaceae	Chanmari, Lohaghat	1650	Roots taken in fever and in internal injury; plant is believed pious and used in several religious ceremonies; increases quantity and quality of milk in livestock.
Cyperus corymbosus Rottboell		Cyperaceae	Bhanargaon, Almora	1350	As fodder.
Cyperus monocephalus Roxb.		Cyperaceae	Chanmari, Lohaghat	1650	Occasionally used as fodder.
Cyperus nutans Vahl		Cyperaceae	Chanmari, Lohaghat	1650	Occasionally used as fodder.
Cyperus rotundus L.	*Motha*	Cyperaceae	Abattmount, Lohaghat	1700	Dried underground parts used in perfumes; plant extract used as diaphoretic and astringent.
Cyrtococcum accrescens (Trinius) Stapf		Poaceae	Pauri	1400	Used as fodder.
D. setigera Roth ex Roemer and Schultes		Poaceae	Motadhang, Pauri	700	Fodder for cattle; helps increase milk quantity and quality.

Contd...

Table 3.1–Contd...

Botanical Name	Vernacular Name (English Name)	Family	Site/ Location	Altitude, m msl	Ethnobotanical Notes
Dactyloctenium aegypticum (L.) P. Beauv.	*Bans* (male Bamboo)	Poaceae	Chanmari, Lohaghat	1650	The plant is used for thatching and roofing, construction, walking sticks, baskets and other containers bows and arrows, mats, fuel, furniture, agricultural implements, sericulture industry as trays for silk worms, religious and cultural rites, musical instruments, etc.; leaves provide good fodder; young buds and rhizome edible; leaf juice rhizome and fruits used in medicine.
Dalbergia lanceolata L. f.	*Takoli, Bitwa*	Fabaceae	Kotdwar	500	Branches and leaves lopped for fodder.
Dalbergia latifolia Roxb.		Fabaceae	Laldhang	600	Foliage for fodder; wood used for furniture and construction purposes; an important tree of social forestry.
Dalbergia sericea G. Don	*Bhandir, Gugar*	Fabaceae	Sara, Dugadda	600	Foliage used as fodder; checks soil erosion.
Dalbergia sissoo Roxb.	*Shisham* (Sissoo)	Fabaceae	Srinagar	1200	Wood commercially well known for construction and furniture; leaves lopped for fodder. The resin used in skin ailments; an important tree of social forestry.
Dalbergia volubilis Roxb.		Fabaceae	Mithiber	1000	Occasional fodder.
Debregeasia longifolia (Burm. f.) Wedd.	*Tusara, Sausaru* (Wild-rhea)	Urticaceae	Mahatgaon, Almora	1450	Plant is good fodder; bark yields strong fibre; plaster made from pulverized bark for bone fracture.

Contd...

Table 3.1–Contd...

Botanical Name	*Vernacular Name (English Name)*	*Family*	*Site/ Location*	*Altitude, m msl*	*Ethnobotanical Notes*
Debregeasia salicifolia (D. Don) Rendle	*Syanru, Tushaira*	Urticaceae	Bhanargaon, Almora	1600	Plant provides good fodder; bark yields strong fibre for ropes and cordages; fruits edible; plaster made from the bark for bone fracture.
Dendrocalamus strictus (Rox.) Nees	*Bans* (Male Bamboo)	Poaceae	Matela, Almora	1200	The plant is used for thatching and roofing, construction, walking sticks, baskets and other containers, bows and arrows, mats, fuel, furniture, agricultural implements, sericulture industry as trays for silk worms, religious and cultural rites, musical instruments, etc.; leaves provide good fodder; young buds and rhizomes edible; leaf juice, rhizome and fruits used in medicine.
Desmodium elegans DC.	*Chamlai*	Fabaceae (Papilionaceae)	Almora	1800	Infusion of roots smelled in epilepsy; roots used as carminatives.
Desmodium laxiflorum DC.		Fabaceae (Papilionaceae)	Almora	1650	Used as fodder.
Desmodium microphyllum (Thunb.) DC.	*Sunsuni*	Fabaceae (Papilionaceae)	Ghingharikhal, Almora	1800	Fodder for cattle.
Desmodium multiflorum DC.		Fabaceae (Papilionaceae)	Ghingharikhal, Almora	1800	As fodder.
Desmodium triflorum (L.) DC.	*Kudalia*	Fabaceae (Papilionaceae)	Ghingharikhal, Almora	1800	Plants browsed by cattle.

Contd...

Table 3.1–Contd...

Botanical Name	*Vernacular Name (English Name)*	*Family*	*Site/ Location*	*Altitude, m msl*	*Ethnobotanical Notes*
Desmostachya bipinnata (L.) Stapf	*Doab, Durva, Dabhena*	Poaceae	Corbette National Park	400	Mats made from the straw; believed to be sacred; considered to be one of the best soil-binding grasses; also browsed by livestock and wild animals.
Deutzia staminea R.Br. ex Wallich	*Ghughtai, Bhatkukri*	Hydrangea-ceae	Gumkhal	1500	Used as fuel as well as fodder; leaves used as a diuretic ; flowers useful in apiculture as bee-forage.
Dichanthium annulatum (Forsk.) Stapf		Poaceae	Abattmount, Lohaghat	1700	Excessively used as fodder; increases milk yields and quality.
Dicliptera bupleuroides Nees	*Kulartore, Kuthhi*	Acanthaceae	Ghingharikhal, Almora	1800	The plant used as fodder; leaf paste applied on wounds to check bleeding; leaf juice useful in cough and gastro-enteritis; flowers as source of bee-forage in apiary.
Dicliptera bupleuroides Nees	*Kulartore, Kuthhi*	Acanthaceae	Patan, Lohaghat	1650	Leaf paste applied on wounds to check bleeding; leaf juice useful in cough and gastro-enteritis; flowers as source of bee-forage in apiary; used as fodder.
Didymocarpus pedicellatus R.Br.	*Pathar-phori, Shila Pushpi*	Gesneriaceae	Kosi, Almora	1250	Leaves used to cure kidney stones.
Digera muricata (L.) Martius	*Kudra, Lesua*	Amarantha-ceae	Matela, Almora	1200	Flowering tops occasionally eaten as vegetable; leaves as fodder.
Digitaria abludens (Roemer and Schultes) Veldkamp	*Ghas*	Poaceae	Matela, Almora	1200	Fodder for livestock.

Contd...

Table 3.1–Contd...

Botanical Name	Vernacular Name (English Name)	Family	Site/ Location	Altitude, m msl	Ethnobotanical Notes
Digitaria adscendens (Kunth) Hennard		Poaceae	Almora	1500	Fodder for cattle.
Digitaria bicornis (Lam.) Roemer and Schultes		Poaceae	Dhikala, Pauri	500	Fodder grass.
Digitaria ciliaris (Retz.) Koeler		Poaceae	Khirsu, Srinagar	1200	Fodder grass.
Digitaria cruciata (Nees ex Steudel) *A. Camus*		Poaceae	Patan, Lohaghat	1650	Fodder grass; helps increase milk quantity and quality.
Digitaria granularis (Trinius) Henrard		Poaceae	Chauri, Lohaghat	1650	Fodder for cattle.
Digitaria stricta Roth ex Roemer and Schultes		Poaceae	Chauri, Lohaghat	1650	Use for fodder.
Dillenia indica L.	*Chalta* (Dillinea)	Dilleniaceae	Matela, Almora	1200	Fruits mixed with honey given in dyspepsia, bronchitis and fever; also edible.
Dioscorea belophylla (Prain) J.O. Voigt ex Haines	*Tairu, Turar* (Wild Yam)	Dioscoreaceae	Dugadda, Srinagar	1200	Leaves provide fodder; tubers edible as well as medicinal, much liked by wildlife.

Contd...

Table 3.1–Contd...

Botanical Name	*Vernacular Name (English Name)*	*Family*	*Site/ Location*	*Altitude, m msl*	*Ethnobotanical Notes*
Dioscorea bulbifera L.	*Genthi* (Potato Yam)	Dioscorea-ceae	Chauri, Lohaghat	1650	Tubers are cooked as vegetable; dried tubers used as an antiseptic, applied on burns and wounds; flowers occasionally visited by bees.
Dioscorea deltoidea Wallich ex Grisebach	*Harvish, Keerish, Tarur*	Dioscorea-ceae	Khirsu	2200	Boiled tubers taken in urinogenital disorders and also used as detergent; occasionally as fodder.
Dioscorea glabra Roxb.	*Tairu*	Dioscorea-ceae	Srinagar	1000	Occasionally as fodder; also used as a medicinal plant.
Dioscorea hispida Dennstaedt		Dioscorea-ceae	Dadmandi	800	Occasionally as fodder plant; also as a medicinal plant.
Dioscorea melanophyma Prain and Burkill	*Mangaii*	Dioscorea ceae	Daira	1500	Occasionally as fodder; used as a medicinal plant also.
Dioscorea pentaphylla L.	*Kou, Phal-alu*	Dioscorea-ceae	Rudraprayag	1500	After thorough boiling and washing, tubers occasionally cooked as vegetable; rarely as fodder.
Dioscorea pubera Blume	*Kasalu*	Dioscorea-ceae	National Corbette Park	600	Occasionally as fodder plant; also used as a medicinal plant.
Diospyros montana Roxb.	*Pinna* (Mottled Ebony)	Ebenaceae	Kosi, Almora	1250	Wood used for agricultural implements; fruits used to intoxicate fishes.
Diospyros tomentosa Roxb.	*Tendu* (Nepal Ebony)	Ebenaceae	Matela, Almora	1200	Fruits edible; wood used for construction and furniture, especially carved articles.

Contd...

Table 3.1–Contd...

Botanical Name	*Vernacular Name (English Name)*	*Family*	*Site/ Location*	*Altitude, m msl*	*Ethnobotanical Notes*
Dodecadenia grandiflora Nees	*Tailiya*	Lauraceae	Bharsar, Binsar	2500	Wood used for agricultural implements; leaves as fodder.
Drymaria cordata (L.) Willd. ex Roemer and Schultes	*Pithpapra*	Caryophyllaceae	Abattmount, Lohaghat	1650	Aqueous paste of the plant given in bile complaints; young leaves cooked as vegetable; the plant is a good fodder.
Echinochloa crus-galli (L.) P. Beauv.	*Jhangora* (Barnyard Millet)	Poaceae	Chanmari, Lohaghat	1650	Use for fodder.
Echinochloa frumentacea Link	*Jhangora* (Barnyard Millet)	Poaceae	Chanmari, Lohaghat	1650	Use for fodder.
Ehretia acuminate R.Br.	*Pudila, nara*	Ehretiaceae	Khanda, narkota	1200	Wood used for various purposes; leaves lopped for fodder. Fruits edible, afresh or pickled; flowers as source of bee-forage, plants useful in social forestry.
Ehretia laevis Roxb.	*Chamror, Khoda*	Ehretiaceae	Almora	1500	Wood used for agricultural implements and for constructions; leaves lopped for fodder; fruits edible; flowers as source of bee-forage; plant useful in social forestry.
Elaeagnus parviflora Wallich ex Royle		Elaeagnaceae	Adhwani, Binsar	2600	Fruits edible, believed to be good in cough and bronchitis; wood used as fuel; leaves lopped for fodder; flowers used as febrifuge and seed powder as an expectorant.

Contd...

Table 3.1–Contd...

Botanical Name	*Vernacular Name (English Name)*	*Family*	*Site/ Location*	*Altitude, m msl*	*Ethnobotanical Notes*
Eleusine compressa (Forsk.) Ascherson and Schweinfurth ex Christensen		Poaceae	Srinagar	1000	Used as fodder.
Eleusine coracana (L.) Gaertner	*Manduwa* (Finger Millet)	Poaceae	Katarmal, Almora	1550	Grains used for edible purposes as substitute of wheat; straw used for fodder; local beverage made from the grains; flour of the grains also used to soften the skin.
Eleusine indica (L.) Gaertner	*Jharnpriya-Kodu* (Crab Grass)	Poaceae	Srinagar	1200	Mainly used as fodder.
Eleusine indica (L.) Gaertner	*Jharnpriya-Kodu* (Crab Grass)	Poaceae	Patan, Lohaghat	1650	Mainly used as fodder.
Eragrostiella nardoides (Trinius) Bor		Poaceae	Sara, Srinagar	1200	Used as fodder.
Eragrostis atrovirens (Desfontaines) Trinius ex Steudel		Poaceae	Chanmari, Lohaghat	1650	Use for fodder.
Eragrostis japonica (Thumb.) Trinius		Poaceae	Mahatgaon, Almora	1450	Use for fodder.

Contd...

Table 3.1–Contd...

Botanical Name	*Vernacular Name (English Name)*	*Family*	*Site/ Location*	*Altitude, m msl*	*Ethnobotanical Notes*
Eragrostis minor Host		Poaceae	Kamleshwar	800	As fodder.
Eragrostis nigra Nees ex Steudel		Poaceae	Khirsu, Dudhạtoli	2200	As fodder.
Eragrostis pilosa (L.) P. Beauv.		Poaceae	Dwarikhal	800	As fodder.
Eragrostis tenella (L.) P. Beauv. ex Roemer and Schultes		Poaceae	Laldhang-Bhabar	600	As fodder.
Eragrostis unioloides (Retz.) Nees ex Steudel		Poaceae	Dwarikhal, Chakisain	1200	As fodder.
Erigeron annuus		Asteraceae	Chanmari, Lohaghat	1650	As fodder.
Erigeron bonariensis L.		Asteraceae	Chauri, Lohaghat	1650	As fodder.
Erigeron canadensis L.		Asteraceae	Patan, Lohaghat	1650	As fodder.
Erigeron karvinskianus DC.		Asteraceae	Khetikhan, Lohaghat	1650	As fodder.

Contd...

Table 3.1–Contd...

Botanical Name	Vernacular Name (English Name)	Family	Site/ Location	Altitude, m msl	Ethnobotanical Notes
Erythrina variegata L.	*Mandar* (Indian Coral Tree)	Fabaceae	Laldhang, Srinagar	800	Bark used as an insecticidal and anthelmintic; plant extract used in snake bite; fruits sometimes used as vegetable; rarely used as fodder.
Eulalia hirtifolia (Hackel) *A. Camus*		Poaceae	Dhumakot-Rewakhet	1400	Fodder for livestock.
Eulalia leschenaultiana (Decne.) Ohwi		Poaceae	Srinagar	1000	Fodder for livestock.
Eulalia mollis (Grisebach) Kuntze		Poaceae	Tarakund-Binsar, Khirsu	2500	Fodder for livestock.
Eulalia quadrinervis (Hackel) Kuntze		Poaceae	Rewakhet, Sundergaon	800	Fodder for livestock.
Eulaliopsis binata (Retz.) Hubbard	*Sabai grass, Bhabar Grass, Babula*	Poaceae	Srinagar, Khandah	1200	Ropes made from the leaves; also as fodder.
Euonymus echinatus Wallich	*Laguliya, Laduli*	Celastraceae	Bharsar	2200	Leaves occasionally as fodder.
Euonymus hamiltonianus Wallich	*Dharua, Dadul*	Celastraceae	Deewakhal	2500	Fuel from wood; fodder from leaves; flowers useful in apiculture as bee forage; useful tree of social forestry.
Euonymus pendulus Wallich	*Bhambeli*	Celastraceae	Binsar, Adhwani	2200	Wood as fuel and agricultural implements; leaves as fodder.

Contd...

Table 3.1–Contd...

Botanical Name	*Vernacular Name (English Name)*	*Family*	*Site/ Location*	*Altitude, m msl*	*Ethnobotanical Notes*
Euonymus tingens Wallich	*Konk, Bhambeli, Kumkum*	Celastraceae	Binsar, Pauri	2400	Wood used for carving; leaves lopped for fodder; bark paste useful in eye diseases. Yellow dye (yellow *chandan*) obtained from the bark, considered sacred by locales and used in various religious ceremonies; flowers useful in apiculture as bee-forage; leaves occasionally used as fodder.
Euonymus tingens Wallich	*Konk, Bhambeli, Kumkum*	Celastraceae	Almora	1800	Wood used for carving; leaves lopped for fodder bark paste useful in eye diseases; yellow dye (yellow *chandan*) obtained from the bark, considered sacred by the locales; used in various religious ceremonies; flowers useful in apiculture as bee-forage.
Eupatorium adenophorum Sprengel	*Kharna, Bakura*	Asteraceae	Abattmount, Lohaghat	1700	An obnoxious weed occupying large grassy localities reducing fodder; crushed leaves applied on wounds.
Euphorbia pilosa L.	*Chuplya, Chounpalu*	Euphorbiaceae	Binsar, Pauri	2400	Decoction of roots given in constipation; fruits emetic; increases milk quantity and quality.
Euphorbia chamaesyce L.	*Dudhi* (Swallow Wort)	Euphorbiaceae	Almora	1650	Plant juice given to infants in constipation and dysentery.
Euphorbia hirta L.	*Dudhi* (Australian Asthma Herb, Garden Spurge)	Euphorbiaceae	Kotdwar, Pauri	1200	Plants used in bronchial infection and asthma; latex is applied on warts. Plants increase milk yield and milk quality.

Contd...

Table 3.1–Contd...

Botanical Name	*Vernacular Name (English Name)*	*Famiiy*	*Site/ Location*	*Altitude, m msl*	*Ethnobotanical Notes*
Euphorbia hypericifolia L.	*Hazardana*	Euphorbia-ceae	Matela, Almora	1200	Leaf infusion given in folk remedy for dysentery, diarrhoea and menorrhagea.
Eurya acuminata DC.	*Bakryal, Deura*	Theaceae	Abattmount, Lohaghat	1700	Wood used as fuel and for agricultural implements; green leaves and young twigs lopped for fodder.
Fagopyrum dibotrys (D.Don) Hara	*Ban Ogal*	Polygona-ceae	Abattmount, Lohaghat	1700	Leaves occasionally used as vegetable; leaf paste rubbed on insect bite.
Fagopyrum esculentum (L.) Moench	*Ougal, Kotu* (Buckwheat)	Polygonaceae	Thailisain, Pauri	1500	Leaves and young branches made into vegetable as well as snacks. Flour of seeds used as a substitute of wheat.
Festuca gigantea (L.) Villars	(Giant , Fescue Tall Brome)	Poaceae	Binsar, Pauri	2500	Used as fodder for livestock; helps increase milk production.
Festuca rubra L.		Poaceae (Creeping Fescue)	Nainidanda-Dhumakot, Pauri	1200	Used as fodder.
Ficus arnottiana (Miq.) Miq.	Paras-Pipal	Moraceae	Dhikala, Pauri	600	Leaves used as fodder.

Contd...

Table 3.1–Contd...

Botanical Name	*Vernacular Name (English Name)*	*Family*	*Site/ Location*	*Altitude, m msl*	*Ethnobotanical Notes*
Ficus auriculata Lour.	Timla, Tirmul	Moraceae	Dugadda, Pauri	1000	Leaves made into cup and plates; also provide good fodder for cattle as well as elephants; ripe fruits edible, unripe fruits made into vegetable; wood as fuel. Plants are conserved for its religious significance and utilization.
Ficus benghalensis L.	*Bargad, Bar* (Banyan Tree)	Moraceae	Matela, Almora	1200	Wood useful for tent poles, yokes, boats, etc.; leaves and twigs good fodder; fruits edible, especially for wildlife; latex medicinal as anti-diabetic; considered as sacred tree.
Ficus clavata Wallich ex Miq.	*Chanchri*	Moraceae	Tarag, Lohaghat	1550	Leaves and branches are used as fodder; fruits edible.
Ficus cunia Buch.-Ham, ex Roxb.	*Khinyo, Khaina, Kharpatra*	Moraceae	Tarag, Lohaghat	1550	Leaves and twigs lopped for fodder; fibre obtained from the bark to prepare ropes; fruits edible; an important agroforestry tree.
Ficus glaberrima Blume	*Khankri*	Moraceae	Dugadda, Pauri	1000	Young twigs and leaves provide good fodder.
Ficus hederacea Roxb.	*Beduli, Laduli*	Moraceae	Binsar, Pauri	1800	Leaves used as fodder.
Ficus hipsida L.f.	*Totmila, Ghogsa*	Moraceae	Dugadda, Srinagar	1000	Young fruits eaten in curries; leaves provide good fodder; bark used as rough fibre.
Ficus laminose Hardwicke	*Gadh-gogsu*	Moraceae	Khoh Valley, Dugadda	600	Leaves provide good fodder.

Contd...

Table 3.1–Contd...

Botanical Name	Vernacular Name (English Name)	Family	Site/ Location	Altitude, m msl	Ethnobotanical Notes
Ficus neriifolia Smith var. *Ficus nemoralis* Wallich ex Miq.	*Dudhla, Thilook*	Moraceae	Tarag, Lohaghat	1550	Leaves used as fodder; fruits edible.
Ficus palmata Forsk.	*Bedu* (Common Fig)	Moraceae	Chanmari, Lohaghat	1550	Leaves and twigs good fodder; fruits delicious in taste, often taken raw with salt or filled inside the bread; fruits medicinal for digestive disorders; useful plant of agroforestry.
Ficus racemosa L.	*Umra, Gular* (Cluster Fig)	Moraceae	Byasghat, Srinagar	900	Fruits edible, raw or cooked; leaves used as fodder; wood useful in religious fires.
Ficus religiosa L.	*Peepal, Pipal* (Peepal)	Moraceae	Kosi, Almora	1200	Leaves good fodder for goats and elephants; figs and bark medicinal, used in bronchitis and skin ailments; regarded as a sacred tree; used in religious rites.
Ficus roxburghii Wallich ex Miq.	*Timla, Tirmul*	Moraceae	Tarag, Lohaghat	1550	Leaves made into cup and plates; also provide good fodder for cattle as well as elephants; ripe fruits edible, unripe fruits made into vegetable; wood as fuel; plants conserved for its religious significance and utilization.
Ficus rumphii Blume	*Kobar, Pilkhi*	Moraceae	Corbette Park	1200	Leaves and branches used as fodder for cattle and elephants. Tree acts as a host for lac-insect.

Contd...

Table 3.1–Contd...

Botanical Name	*Vernacular Name (English Name)*	*Family*	*Site/ Location*	*Altitude, m msl*	*Ethnobotanical Notes*
Ficus sarmentosa Buch.-Ham. ex J.E. Smith	*Beduli*	Moraceae	Thailisain, Pauri	2000	Fruits edible; leaves provide good fodder.
Ficus semicordata Buch.-Ham. ex J.E. Smith	*Khaina, Jharphali*	Moraceae	Jhair, Pabau, Pauri	1400	Leaves and twigs lopped for fodder; fibre obtained from the bark to prepare ropes; fruits edible; an important agroforestry tree.
Flacourtia indica (Burm. f.) Merrill	*Kandai, Kangu*	Flacourtia-ceae	Chipalghat, Laldhang	1200	Fruits edible; leaves as fodder; decoction of fruits given in hepatitis, fever and diarrhoea; leaves and bark paste applied on wounds; flowers as a source of bee-forage.
Flemingia bracteata (Roxb.) Wight	*Veer-brat, Shal-Parni*	Fabaceae	Kosi, Almora	1200	Fodder for cattle; plant extract used in bronchial diseases.
Flemingia macro-phylla (Willd.) Prain ex Merrill		Fabaceae	Srinagar	1200	Leaves used as fodder; root paste externally applied on swellings and ulers.
Flemingia paniculata Wallich ex Benth.		Fabaceae	Laldhang	800	Occasionally used as fodder.
Flemingia procum-bens Roxb.	*Chauna, Cheena*	Fabaceae	Katarmal, Almora	1600	Underground tubers edible; paste of roots with curd given in stomach disorders.
Flemingia semialata Roxb. ex Aiton f.		Fabaceae	Abattmount, Lohaghat	1700	

Contd...

Table 3.1–Contd...

Botanical Name	*Vernacular Name (English Name)*	*Family*	*Site/ Location*	*Altitude, m msl*	*Ethnobotanical Notes*
Flemingia strobilifera (L.) R.Br.	*Bari-Kapasi, Kausunt*	Fabaceae	Abattmount, Lohaghat	1700	Plants used as brooms. Leaves medicinal in hysteria; also used as fodder.
Fraxinus micrantha Lingelsheim	*Angu, Angar* (Himalayan Ash)	Oleaceae	Khetikhan, Lohaghat	1650	Wood hard and close grained, used for walking sticks and agricultural implements; leaves and bark extract used in fever and blood dysentery.
Galinsoga parviflora Cav.	*Marchya* (Gallant Soldier)	Asteraceae	Chauri, Lohaghat	1650	Plant as fodder; plant rubbed on the skin as an antidote of nettle stings.
Galium aparine L.	*Khulkusa, Kuri*	Rubiaceae	Katarmal, Almora	1550	Extract of leaves used as astringent; plant paste applied on skin diseases.
Galium elegans Wallich	*Kutub, Manjeethee*	Rubiaceae	Almora	1600	Plant extract given in colic, dispepsis, as well as in jaundice.
Galium rotundifolium L.	*Kutub, Manjeethee*	Rubiaceae	Abattmount, Lohaghat	1700	Plant extract given in colic, dyspepsia, as well as in jaundice.
Garnotia tenella (Arn. ex Miq.) Janowsky		Poaceae	Dugadda, Lansdowne	1500	Used for thatching, rarely as fodder.
Garuga pinnata Roxb.	*Titmar, Kaikar*	Burseraceae	Matela, Almorabark	1200	Wood has high timber value; leaves used for fodder; yields tannin; flowers useful source of bee-forage in apiculture; an important tree of social forestry.

Contd...

Table 3.1–Contd...

Botanical Name	Vernacular Name (English Name)	Family	Site/ Location	Altitude, m msl	Ethnobotanical Notes
Geranium wallichianum D. Don ex Sweet	*Ratanjot, Kaphlya, Laljari* (Robert Geranium)	Geraniaceae	Abattmount, Lohaghat	1700	Root juice is used in otorrhoea and ophthalmia; red dye obtained from the roots said to promote growth of hairs, as well as to dye woolen clothes.
Girardinia diversifolia (Link) Friis	*Bhainsya-Kandali* (Himalayan Nettle)	Urticaceae	Katarmal, Almora	1800	Stem yields fibre for ropes or cordages; leaf juice given in gonorrhoea.
Glycine max (L.) Merrill	*Bhat* (Soybean)	Fabaceae (Papilionaceae)	Mahatgaon, Almora	1600	Seeds rich source of proteins, fats; used as pulse and oil-yielding resource.
Gmelina arborea Roxb.	*Gumbahar, Kumbhar, Shivparni* (Malaya-Bushbeech)	Verbenaceae	Almora	1500	Wood produces valuable timber for furniture, carving and craft work; leaves used as an anti-septic and flower as blood purifier; roots highly medicinal and tonic, as one of the ingredients of 'Dasmula' of ayurveda; flowers as source of bee-forage in apiary; useful plant for social forestry.
Gomphrena celosioides Martius		Amaranthaceae	Kosi, Almora	1250	Browsed by animals.
Grewia asiatica L.	*Pharsula, Bhyunl, Dhaman*	Tiliaceae	Dugadda	800	Inner bark yields strong fibre; leaves lopped for fodder; ripe fruits edible; an important agroforestry tree.

Contd...

Table 3.1–Contd...

Botanical Name	*Vernacular Name (English Name)*	*Family*	*Site/ Location*	*Altitude, m msl*	*Ethnobotanical Notes*
Grewia eriocarpa A.L. Juss.	*Pharasain, Bhimla, Phalsa*	Tiliaceae	Dugadda	800	Inner bark yields strong fibre; leaves lopped for fodder; wood for agricultural implements; an important agroforestry tree.
Grewia optiva J.R. Drummond ex Burret	*Bhimal*	Tiliaceae	Tarag, Lohaghat	1650	Bark fibre extensively used for ropes, nets, sacs, brushes, brooms, etc.; sticks after peeling off the bark used to lit fire; leaves provide good fodder; fruits edible and medicinal; a very important agroforestry tree.
Grewia sapida Roxb. ex DC.		Tiliaceae	Corbette National Park	600	Inner bark yields strong fibre; leaves lopped for fodder.
Grewia sclerophylla Roxb. ex D. Don	*Phrsia, Garhbheli, Pharsia*	Tiliaceae	Corbette National Park	600	Stem yields fibre suitable for ropes; leaves lopped for fodder; ripe fruits edible.
Grewia serrulata DC.	*Gar-bhimali, Kath-bhimla*	Tiliaceae	Laldhang, Kotdwar	400	Lopped for fodder; stem yields strong fibre; fruits edible; an important agroforestry tree.
Gymnostemma pedata Blume		Cucurbitaceae	Chauri, Lohaghat	1650	Browsed by animals.
Hackelochloa granularis (L.) Kuntze	*Kangani, Trinpali*	Poaceae	Sara, Srinagar	1200	Fodder for livestock.
Hedera nepalensis K. Koch	*Laguli, Mithiari* (Ivoy, Creeping Ivoy)	Araliaceae	Chelusain, Chirogi	2300	Leaf and fruit paste applied on ulcers; leaf juice given in dyspepsia.

Contd...

Table 3.1–Contd...

Botanical Name	*Vernacular Name (English Name)*	*Family*	*Site/ Location*	*Altitude, m msl*	*Ethnobotanical Notes*
Hedychium spicatum Buch.-Ham. ex J.E. Smith	*Bagldu, Ban-Haldi, Kachor* (Spiked Ginger Lily)	Zingibera-ceae	Abattmount, Lohaghat	1700	Boiled underground parts eaten with salt; roasted powder of rhizome given in asthma; seeds believed to cause abortion; decoction of rhizome with saw-dust of Deodara taken in tuberculosis.
Helictotrichon virescens (Nees ex Steudel) Henrard		Poaceae	Dudhatoli	2400	Browsed by aniamls.
Hemarthria compressa (L.f.) R.Br.		Poaceae	Srinagar, Laldhang	1000	Occasionally used as fodder.
Heteropogon contortus (L.) P. Beauv. ex Roemer and Schultes	*Kumrya-ghas* (Spear grass, Bellary Grass)	Poaceae	Mahatgaon, Almora	1450	Grass stores as hay for scarcity period.
Hibiscus cannabinus L.	*Patsan* (Kenaf, Bimli)	Malvaceae	Srinagar	800	Stem yields a strong fibre; leaves and flowers medicinal; tender twigs as fodder.
Hippophae salicifolia D. Don	*Chuk, Tarwa, Dharchuk* (Sallow Thorn, Seabuckthorn)	Elaeagna-ceae	Bharsar	2200	Fruits eaten raw or made into various preparations, also used in local beverages; syrup of immature fruits used to remove dandruff of hairs; flowers useful source of bee-forage; leaves rarely used as fodder.

Contd...

Table 3.1–Contd...

Botanical Name	*Vernacular Name (English Name)*	*Family*	*Site/ Location*	*Altitude, m msl*	*Ethnobotanical Notes*
Holoptelea integrifolia (Roxb.) Planchon	*Papri, Banchilla* (Indian Elm)	Ulmaceae	Kosi, Almora	1250	Fruits much likened by monkeys; wood used for construction, charcoal and fuel; decoction of bark applied on rheumatic pain; an important tree of social forestry.
Hymenodictyon flaccidum Wallich	*Pinchwa*	Rubiaceae	Kalyasaur, Pauri	1200	Leaves as fodder; wood for agricultural implements.
Hymenodictyon orixense (Roxb.) Mabberley var. *Hymenodictyon excelsum* (Roxb.) Wallich	*Boinga*	Rubiaceae	Dugadda, Kotdwar	800	Wood used for agricultural implements; bark for tanning and as febrifuge; leaves as fodder.
Hypericum choisianum Wallich ex N. Robson var. *H. hookerianum* Wight and Arn.	*Phulya, Basanti*	Hyperica-ceae	Khirsu	2000	Leaves and branches as fodder; leaf powder used in tertiary fever.
Hypericum elodeoides Choisy	*Basanti*	Hyperica-ceae	Khirsu, Dudhatoli	2400	Sometimes leaves used as fodder.
Hypericum japonicum Thunb. ex Murray		Hyperica-ceae	Baijraun, Pauri	2300	Sometimes leaves used as fodder.

Contd...

Table 3.1–Contd...

Botanical Name	*Vernacular Name (English Name)*	*Family*	*Site/ Location*	*Altitude, m msl*	*Ethnobotanical Notes*
Hypericum oblongifolium Choisy	*Chitroi, Chaya, Basant*	Hypericaceae	Binsar, Pauri	2300	Leaves lopped for fodder; roots yield a yellow dye; decoction of leaves and stem given to facilitate delivery.
Hypericum perforatum L.	*Choli-phulya, Basant* (Common St. John's Wort, Kalamath Weed)	Hypericaceae	Binsar, Pauri	2000	Stem as fuel; leaves as fodder; paste of leaves applied on cuts to check bleeding. Infusion of leaves given in malarial fever.
Hypericum uralum Buch.-Ham. ex D. Don	*Bhyoul, Tumbhul*	Hypericaceae	Khirsu	2000	Seed powder against food poisoning and as an abortifacient. Sometimes leaves used as fodder.
Hypericum wightianum Wallich ex Wight and Arn.		Hypericaceae	Binsar	2500	Sometimes leaves lopped for fodder.
Ilex dipyrena Wallich	*Kandara, Kanel*	Aquifoliaceae	Khirsu	2200	Wood for agricultural implements and as fuel; leaves rarely as fodder.
Ilex excelsa (Wallich) Hook. f.	*Gauloo, Gholya*	Aquifoliaceae	Singtali	1500	Wood used as fuel; leaves as fodder; flowers useful in apiculture as bee-forage.
Ilex odorata Buch.-Ham. ex D. Don		Aquifoliaceae	Rudraprayag	1500	Wood used as fuel; leaves as fodder.
Impatiens scabrida DC.	*Ban-til, Tilua*	Balsaminaceae	Abattmount, Lohaghat	1700	Seeds and seed oil edible; mucilaginous substance of the stem believed to cause abortion; occasionally also used as fodder.

Contd...

Table 3.1–Contd...

Botanical Name	Vernacular Name (English Name)	Family	Site/ Location	Altitude, m msl	Ethnobotanical Notes
Imperata cylindrica (L.) P. Beauv.	*Sauraun, Sirau* (Thatch Grass)	Poaceae	Abattmount, Lohaghat	1700	Woody hairs of the inflorescence used for staunching the wounds and for stuffing purposes; roots used as tonic.
Indigofera cassioides Rottler ex DC.	*Sakena*	Fabceae	Katarmal, Almora	1600	Flowers made into vegetables and snacks, often filled in between breads; leaf powder taken in malarial fever; foliage used as fodder.
Indigofera dosua Buch.-Ham. ex D.Don		Fabaceae	Abattmount, Lohaghat	1700	Occasionally used as fodder.
Indigofera heterantha Wallich ex Brandis	*Sakina, Kathi, Kathoj* (Himalayan Indigo)	Fabaceae	Kosi, Almora	1250	Flowers used as vegetable; leaves lopped for fodder; twigs made into baskets or containers; leaf juice taken in diarrhoea, dysentery and cough.
Ipomoea muricata (L.) Jacquin	*Michai*	Convolvula-ceae	Tarag, Lohaghat	1650	Roasted seed-powder with water given in fever and constipation.
Isachne albens Trinius		Poaceae	Abattmount, Lohaghat	1700	Use for fodder.
Ischaemum rugosum Salisbury	*Maronda, Murchi*	Poaceae	Kotdwar, Kalaghati	400	Fodder grass.

Contd...

Table 3.1–Contd...

Botanical Name	*Vernacular Name (English Name)*	*Family*	*Site/ Location*	*Altitude, m msl*	*Ethnobotanical Notes*
Justicia simplex (D. Don) Yamazaki		Acanthaceae	Abattmount, Lohaghat	1700	Occasionally as fodder.
Koeleria macrantha (Ledebour) Schultes	*Crested Hair Grass*	Poaceae	Sara-Paukhal	1500	Used as fodder.
Kydia calycina Roxb.	*Phuilau Pulao*	Malvaceae	Kosi, Almora	1200	Bark yields a coarse fibre; wood used for various construction purposes; leaves lopped for fodder; bark paste applied on body swellings.
Lactuca dissecta D.Don		Asteraceae	Majkhali, Almora	2250	Occasionally used as fodder.
Lactuca dissecta D.Don		Asteraceae	Chaubatiya, Almora	2250	Occasionally as fodder.
Lactuca dolichophylla Kitamura		Asteraceae	Chaubatiya, Almora	2250	Occasonally as fodder.
Laggera alata (D.Don) Schultz-Bipontinus ex Oliver	*Lumra*	Asteraceae	Chaubatiya, Almora	2250	Plants as fodder; heads used in psycho medicines.
Lannea coromandelica (Houttuyn) Merrill	*Kalmina* (Wodier)	Anacardiaceae	Mahatgaon, Almora	1500	Bark for tanning; fuel and agricultural implements from wood; fodder from leaves; gum and bark given in diarrhoea; flowers useful source of bee-forage; fruit powder used to intoxicate fishes.

Contd...

Table 3.1–Contd...

Botanical Name	*Vernacular Name (English Name)*	*Family*	*Site/ Location*	*Altitude, m msl*	*Ethnobotanical Notes*
Lathyrus aphaca L.	*Kurphali* (Yellow Vetehling)	Fabaceae	Srinagar	1200	Used as fodder.
Lathyrus sativus L.	*Latri, Tiuri* (Grass Pea, Chickling Pea)	Fabaceae	Srinagar	1200	Used as fodder.
Lathyrus sphaericus Retz.		Fabaceae	Dugadda	1000	Whole plant used as fodder.
Leptochloa chinensis (L.) Nees		Poaceae	Rasulpur-Laldhang	400	Fodder Grass.
Leptochloa panacea (Retz.) Ohwi		Poaceae	Kamleshwar-Srinagar	1200	Preferred by cattle.
Leptodermis lanceolata Wallich	*Padera, Padar*	Rubiaceae	Abattmount, Lohaghat	1700	Occasionally browsed by cattle; bark paste externally applied in migraine; flowers as an important source of bee-forage in apiculture.
Lespedeza gerardiana Graham ex Maxim.	*Khunja*	Fabaceae	Chaubatiya, Almora	2250	Flowers extract as blood purifier; twigs made into rough brooms to collect food grains.
Lespedeza juncea (L.f.) Persoon		Fabaceae	Chaubatiya, Almora	2250	

Contd...

Table 3.1–Contd...

Botanical Name	*Vernacular Name (English Name)*	*Family*	*Site/ Location*	*Altitude, m msl*	*Ethnobotanical Notes*
Leucas indica (L.) R.Br. ex Vatke		Laminaceae	Tarag, Lohaghat	1650	
Leucas lanata Benth.	*Bis-kapra*	Lamiaceae	Chauri, Lohaghat	1650	Plant infusion given with honey in the treatment of whooping cough; young shoots cooked as vegetable.
Ligustrum compactum (Wallich ex DC.) Hook. f. and Thomson ex Brandis	*Kakurchha*	Oleaceae	Katarmal, Almora	1750	Leaves used as fodder.
Ligustrum indicum (Lour.) Merrill	*Mercha* (Indian Privet)	Oleaceae	Katarmal	1750	Leaves lopped for fodder; wood for minor articles.
Lindenbergia indica (L.) Vatke	*Makria-jhar, Phiunl*	Scrophulariaceae	Abattmount, Lohaghat	1700	Leaves used in bronchitis, poultice applied on cuts and wounds.
Litsea elongate (Nees) Hook. f.		Lauracea	Mahatgaon, Almora	1450	Wood used for making agricultural implements.
Litsea glutinosa (Lour.) Robinson	*Singau* (Common tallow, Laurel)	Lauraceae	Kosi, Almora	1200	Wood used for agricultural implements; plaster made from the bark applied on fractured bones.

Contd...

Table 3.1–Contd...

Botanical Name	*Vernacular Name (English Name)*	*Family*	*Site/ Location*	*Altitude, m msl*	*Ethnobotanical Notes*
Litsea monopetala (Roxb.) Persoon	*Katmara, Karka, Katmala*	Lauraceae	Matela, Almora	1200	Leaves chiefly used for silk-worm farming, occasionally as fodder.
Lolium temulentum L.	*Mochni* (Darnel, Rye Grass)	Poaceae	Srinagar	1000	Fodder Grass.
Lonicera quinquel-ocularis Hardwicke	*Taknoi, Bad-kukura, Chamaksin* (Himalayan Honey Suckle)	Caprifoliaceae	Abattmount, Lohaghat	1700	Fruits edible; stem used for walking sticks; flowers useful source of bee-forage in apiary.
Lotus corniculata L.		Fabaceae	Binsar	2600	Grazed by cattle; good soil-binder.
Machilus duthiei King ex Hook. f.	*Kaul, Kaula, Sairi, Bhadrao*	Lauraceae	Tarag, Lohaghat	1650	Wood used for construction and fuel; leaves and young shoots lopped for fodder.
Madhuca longifolia (Koenig) Mac Bride	*Mahwa* (Butter Tree)	Sapotaceae	Kosi, Almora	1250	Wood for various constructions; flowers and fruits edible afresh or made into several preparations; useful in apiculture as bee-forage; seeds yield an oil; used for candle and soap manufacturing; flowers often used in local beverages.
Maoutia puya (Hook.) Wedd.	*Phur-khagsa, Safed Khagsa*	Urticaceae	Mahatgaon, Almora	1400	Leaves lopped for fodder; bark yields strong fibre, used for fishing-nets and ropes.

Contd...

Table 3.1–Contd...

Botanical Name	Vernacular Name (English Name)	Family	Site/ Location	Altitude, m msl	Ethnobotanical Notes
Maoutia puya (Hook.) Wedd.	*Phur-khagsa, Safed-khagsa, Puya*	Urticaceae	Srinagar	1200	Leaves lopped for fodder; bark yields strong fibre, used for fishing-nets and ropes.
Medicago lupilina L.		Fabaceae	Matela, Almora	1200	As fodder, as well as fertilizer.
Medicago polymorpha L.	(Tooth Clover, Tooth Bur Clover)	Fabaceae	Matela, Almora	1200	Leaves rarely used as vegetable; also as green manure.
Melia azedarach L.	*Bakain, Denkan* (Barbedos Lilac)	Meliaceae	Kosi, Almora	1300	Leaves, fruits and seeds are useful in skin diseases as well as in rheumatic pains; fruits and leaves as antiseptic and wormicides; fruit pulp regarded as an abortifacient; leaves as fodder; wood useful for household and agricultural implements; flowers useful source of bee-forage in apiary; important tree of social forestry.
Melilotus alba Medikus ex Desrou-sseaux	*Safed Senji, Safed Murela* (White Sweet Clover, White Melio)	Fabaceae	Matela, Almora	1250	As fodder and fertilizer.
Melilotus indica (L.) Allioni	*Ban Methi* (Yellow Sweet Clover Flower)	Fabaceae	Mahatgaon, Almora	1500	Roasted seeds given in diarrhoea and dysentery; plant as fodder.

Contd...

Table 3.1–Contd...

Botanical Name	*Vernacular Name (English Name)*	*Family*	*Site/ Location*	*Altitude, m msl*	*Ethnobotanical Notes*
Meliosma dillenii-folia (Wallich ex Wight and Arn.) Walpers	*Goguna, Gulpha*	Sabiaceae	Kundinagar-Binsar	2600	Wood used as fuel, leaves as fodder.
Microstegium falconeri (Hook. f.) Clayton		Poaceae	Dwarikhal	1600	Grazed by animals.
Microstegium nudum (Trinius) *A. Camus*		Poaceae	Sara	800	Grazed by aniamls.
Microtyloma sar-garhwalensis Gaur and Dangwal	*Jangli Gahath* (Wild Horse Gram)	Fabaceae	Sara-Garhwal	1800	Browsed by animals.
Miliusa velutina (Dunal) Hook. f. and Thomsom	*Domsal*	Magnoliaceae	Dugadda	700	Wood used for agricultural implements; leaves as fodder.
Milletia extensa (Benth.) Baker	*Gauja, Gauj, Gonj*	Fabaceae	Dugadda	1000	Branches yield coarse fibre for ropes and cordages; leaves lopped for fodder. Root paste mixed with sesame oil; applied on cutaneous eruptions; plants suitable to landslide zones.

Contd...

Table 3.1–Contd...

Botanical Name	*Vernacular Name (English Name)*	*Family*	*Site/ Location*	*Altitude, m msl*	*Ethnobotanical Notes*
Milletia extensa (Benth.) Baker	*Gauju, Gauj, Gonj*	Fabaceae	Bahanargaon, Almora	1600	Branches yield coarse fibre for ropes and cordages; leaves lopped for fodder; root paste mixed with sesame oil, applied on cutaneous eruptions; plants suitable to landslide zones.
Mimosa himalayana Gamble	*Kingrei, Aila*	Mimosaceae	Almora	1600	Leaves as fodder of sheep and goats; gun powder charcoal made from the wood; leaves medicinal used for cough and cold, bronchitis and urinary complaints.
Miscanthus nepalensis (Trinius) Hackel		Poaceae	Abattmount, Lohaghat	1700	Used for fodder.
Mitragyna parviflora (Roxb.) Korthals	*Phaldu, Kadam, Kaim*	Rubiaceae	Laldhang, Kotdwar	800	Wood used for building, furniture, agricultural implements; and carved articles as cup, plates, etc.; leaves as fodder; bark as fibre; roasted seeds used as purgative; an important tree of avenue.
Mnesithea laevis (Retz.) Kunth		Poaceae	Srinagar	1200	Occasionally as fodder.
Morus alba L.	*Tooth, Tutri* (White Mulberry)	Moraceae	Tarag, Lohaghat	1650	Leaves feed silkworms and fodder to livestock; fruits edible; wood used for agricultural tools and implements and other articles.
Morus macroura Miq.	*Toot, Shehtoot*	Moraceae	Matela, Almora	1200	Fruits edible; leaves used as fodder; wood used for making furniture and other articles; useful tree of social forestry.

Contd...

Table 3.1–Contd...

Botanical Name	Vernacular Name (English Name)	Family	Site/ Location	Altitude, m msl	Ethnobotanical Notes
Morus serrata Roxb.	*Keemu, Shehtoot* (Himalayan Mulberry)	Moraceae	Matela, Almora	1200	Leaves lopped for fodder; fruits edible; wood used for cabinet works and other articles.
Mucuna nigricans (Lour.) Steudel	*Bhaisalu, Gaunchhi*	Fabaceae	Kosi, Almora	1250	Leaves boiled with *rajma* beans, supposed to dissolve renal calculi; roasted seeds edible and also applied on piles; root decoction given in cholera.
Mucuna pruriens (L.) DC.	*Dankuli-chemi, Gaunch*	Fabaceae	Matela, Almora	1200	Twigs and leaves lopped for fodder.
Muhlenbergia huegelii Trinius		Poaceae	Chauri, Lohaghat	1650	Use for fodder.
Myrsine africana L.	*Chupra, Pahari-cha*	Myrsinaceae	Abattmount, Lohaghat	1700	Fruit used as laxative in colic, sometimes anthelmintic, traditionally with the name of "Bayabirang"; flowers useful in apiculture as bee-forage.
Myrsine semi-serrata Wallich	*Gaunta, Bains*	Myrsinaceae	Sara, Pauri	2000	Wood used for agricultural implements; fruit laxative, used in colic; leaves occasionally used for fodder.
Neolitsea cuipala (Buch.-Ham. ex D. Don) Kostermans	*Lampatiya*	Lauraceae	Binsar	2300	Twigs used as fuel and leaves as fodder. Fruit paste with camphor applied on infested wounds of cattle. Wood used for agricultural implements.

Contd...

Table 3.1–Contd...

Botanical Name	*Vernacular Name (English Name)*	*Family*	*Site/ Location*	*Altitude, m msl*	*Ethnobotanical Notes*
Neyraudia arundinacea (L.) Hennard	*Naltura, Bicchhru*	Poaceae	Sara, Srinagar	1000	Widely used for brooms, also as a good fodder.
Nicandra physalodes (L.) Gaertner	*Tambukya, Endu* (Apple of Peru)	Solanaceae	Abattmount, Lohaghat	1700	Plant used as an anthelmintic and insecticidal; leaf extract locally used to kill lice; believed to be poisonous for cattle.
Nitsea monopetala (Roxb.) Persoon	*Katmara, Karka, Katmala, Meda*	Lauraceae	Srinagar	1200	Leaves chiefly used for silk-worm farming, occasionally as fodder.
Olea glandulifera Wallich ex G. Don	*Gair, Gaid, Galdu* (Indian Olive)	Oleaceae	Almora	1700	Reddish grey wood used in construction, turnery and agricultural implements; leaves lopped for fodder; decoction of ark used in anti-periodic fever; flowers as source of bee-forage; an important tree of social forestry.
Oplismenus burmannii (Retz.) *P. Beauv.*		Poaceae	Corbette National Park	600	Used as fodder.
Oplismenus compositus (L.) P. Beauv.	*Ghoria Ghas*	Poaceae	Abattmount, Lohaghat	1700	Use for fodder.
Oplismenus undulatifolius (Ardoino) P. Beauv.		Poaceae	Chakiasain, Thailisain	2300	Grazed by animals.

Contd...

Table 3.1–Contd...

Botanical Name	Vernacular Name (English Name)	Family	Site/ Location	Altitude, m msl	Ethnobotanical Notes
Orecocnide frutescens (Thumb.) Miq.	*Sianru*	Urticaceae	Dugadda	1000	Fibre used for ropes and nets; fruits are edible; leaves provide fodder.
Origanum vulgare L.	*Bantulsi, Jonkjari* (Common Marjoram)	Lamiaceae	Abattmount, Lohaghat	1700	Plant extract used in bronchitis, colic and diarrhoea; leaves eaten as vegetable and flavouring agent; flowering branches hung on the houses to get rid of bad spirits.
Orobanche aegyptiaca Persoon	*Tokra, sarsanbanda* (Aegptian-Broomrape)	Orobanchaceae	Rishikesh	400	Grazed and browsed by cattle and goats.
Oropetium thomaeum (L.f.) Trinius		Poaceae	Chauras-Srinagar	1000	Occasionally serves as fodder.
Ougeinia dalbergioides Benth.	*Saujan, Sandan*	Fabaceae	Tarag, Lohaghat	1650	Twigs lopped for fodder; timber used for various purposes, particularly for making vessels; bark extract used to intoxicate fishes; gum used in digestive troubles; quite useful for bee-forage and soil conservation; an important tree of social forestry and also acts as soil binder.
Oxalis corniculata L.	*Bhilmori* (Indian Sorrel)	Oxalidaceae	Patan, Lohaghat	1650	Leaves taken as salad or cooked as vegetable; leaf juice dropped in cataract and conjunctivitis.
Oxalis dehradunensis Raizada	*Khatura*	Oxalidaceae	Patan. Lohaghat	1650	

Contd...

Table 3.1–Contd...

Botanical Name	*Vernacular Name (English Name)*	*Family*	*Site/ Location*	*Altitude, m msl*	*Ethnobotanical Notes*
Panicum antidotale Retz.	*Hadjodi* (Blue Panic)	Poaceae	Mahatgaon, Almora	1450	As fodder; good soil binder
Panicum curviflorum Hornem.		Poaceae	Srinagar	1000	Occasionally as fodder.
Panicum miliaceum L.	*Cheena* (Hog Millet, Common Millet)	Poaceae	Chauri, Lohaghat	1650	Grains boiled and cooked as rice and flour used as bread; grains also used in local beverages; widely used as fodder.
Panicum paludosum Roxb.		Poaceae	Gumkhal	1600	Occasionally as fodder.
Panicum psilopodium Trinius		Poaceae	Abattmount, Lohaghat	1700	Use for fodder.
Parthenocissus semicordata (Wallich) Planchon	*Bhunera, Laguli*	Vitaceae	Abattmount, Lohaghat	1700	Fruits edible, leaves as fodder.
Paspalidium flavidum (Retz.) Camus		Poaceae	Patan, Lohaghat	1650	Use for fodder.
Paspalum paspalodes (Michaux) Scribn.		Poaceae	Patan, Lohaghat	1650	Use for fodder.

Contd...

Table 3.1–Contd...

Botanical Name	*Vernacular Name (English Name)*	*Family*	*Site/ Location*	*Altitude, m msl*	*Ethnobotanical Notes*
Paspalum scrobiculatum L.		Poaceae	Pauri, Srinagar	1400	Root paste applied around the belly in painful urination; extract dropped in eye ailments; poisonous grass for cattle and human beings.
Passiflora caerulea L.	*Gharibel* (Passiflower)	Passiflora-ceae	Abattmount, Lohaghat	1650	Seeds used as a tonic in general debility; leaves are smeared and inhaled in cold and cough.
Pavetta indica L.	*Angari* (Indian Pellet Shrub)	Rubiaceae	Dugadda, Kalyasaur	1000	Leaves used as fodder; roots and leaves medicinal, used in jaundice and dropsy; flowers as source of bee-forage.
Pennisetum flaccidum Grisebach	*Phark, Phirki*	Poaceae	Chauri, Lohaghat	1650	Grazed by cattle; as a nutritious grass usually conserved along the border of fields.
Pennisetum orientale Richard		Poaceae	Chauri, Lohaghat	1500	Used as fodder.
Pennisetum typhoides Burman	*Bajra*	Poaceae	Chauri, Lohaghat	1650	Grazed by cattle, as a nutritious grass usually conserved along the border of fields.
Perotis hordeiformis Nees ex Hook and Arn.		Poaceae	Laldhang, Chakiasain	600	Sometimes used as fodder.
Persea duthiei (King ex Hook. f.) Kostermans	*Kaula, Sairi, Bhadrao*	Lauraceae	Bharsar, Binsar	2200	Wood used for construction and fuel; leaves and young shoots lopped for fodder.

Contd...

Table 3.1–Contd...

Botanical Name	*Vernacular Name (English Name)*	*Family*	*Site/ Location*	*Altitude, m msl*	*Ethnobotanical Notes*
Phacelurus speciosus (Steudel) C.E. Hubbard	*Ainma*	Poaceae	Binsar	2400	Infusion of plant taken as anthelmintic to kill the intestinal worms; good soil-binder grass; occasionally used as fodder.
Phalaris minor Retz.	*Phulla* (Small Canary Grass)	Poaceae	Mahatgaon, Almora	1500	Root paste applied on wounds.
Phoebe laceolata (Nees) Nees	*Kaula, Kekra, Haulia*	Lauraceae	Laldhang, Chipalghat	1200	Wood for cabinet works; leaves as fodder. Berries used against wounds and sores.
Phragmites karka (Retz.) Trinius ex Steudel	*Khailwa, Narkul, Natkat* (Reed)	Poaceae	Corbette National Park, Kotdwar	400	Used for thatching; occasionally as fodder.
Phyllanthus parvifolius Buch.-Ham. ex D. Don	*Lalbhuin, Serhi*	Euphorbiaceae	Bharsar, Deewakhal	2000	Leaves occasionally used as fodder.
Phyllodium pulchellum (L.) Desvaux		Fabaceae	Dadamandi	1000	As fodder.
Pimpinella achilleifolia (DC.) C.B. Clarke		Apiaceae	Patan, Lohaghat	1650	Occasionally as fodder.

Contd...

Table 3.1–Contd...

Botanical Name	*Vernacular Name (English Name)*	*Family*	*Site/ Location*	*Altitude, m msl*	*Ethnobotanical Notes*
Pimpinella acuminata (Edgew.) C.B. Clarke	*Raulee*	Apiaceae	Abattmount, Lohaghat	1700	Plant extract with curd taken in diarrhea and dysentery; occasionally cooked as vegetable.
Pimpinella diversifolia DC.	*Teroi, Phoree*	Apiaceae	Abattmount, Lohaghat	1700	Plant extract given in digestive disorders as well as in cold and cough.
Pistacia integerrima Stewart	*Kokra-Singi, Kakroi*	Sapindaceae	Abattmount, Lohaghat	1700	Wood used for local crafts and toys; the horn shaped galls (*kakra-singi*) found on the leaves and petioles used for dyeing and tanning and also in the indigenous medicines to relieve cough and fever.
Plantago lanceolata L.	*Luhuriya, Isabgol* (Ribwort Plantain)	Plantaginaceae	Abattmount, Lohaghat	1700	Paste from leaves and seeds applied on cuts, wounds and piles; plant parts tied around the belly of infants for good health; flowers occasionally visited by bees.
Plectranthus japonicus (Burm. f.) Koidz		Lamiaceae	Abattmount, Lohaghat	1700	
Poa annua L.	*Ghas, Bug* (Annual Meadow Grass)	Poaceae	Chauri, Lohaghat	1650	Grazed by animals.
Poa nemoralis L.	*Wood Meadow Grass*	Poaceae	Pauri, Binsar	2400	Grazed by animals.
Poa pratensis L.		Poaceae	Patan, Lohaghat	1850	Grazed by animals.

Contd...

Table 3.1–Contd...

Botanical Name	*Vernacular Name (English Name)*	*Family*	*Site/ Location*	*Altitude, m msl*	*Ethnobotanical Notes*
Pogonatherum crinitum (Thunb.)		Poaceae	Srinagar	1200	Occasionally as fodder.
Pogonatherum paniceum (Lam.) Hackel	*Ringalia Ghas*	Poaceae	Binsar	2200	Occasionally as fodder.
Pollinia quad-rinervis Hackel		Poaceae	Patan, Lohaghat	1650	Grazed by animals.
Polygonum capitatum Buch.-Ham. ex D.Don	*Kafalya*	Polygona-ceae	Patan, Lohaghat	1650	
Polygonum hydropiper L.	*Ameta* (Water Pepper)	Polygona-ceae	Tarag, Lohaghat	1650	The plant contains acrid juice, used as a diuretic agent.
Polypogon fugax Nees ex Steudel		Poaceae	Srinagar	1200	Liked by cattle as fodder.
Polypogon monspeliensis (L.) Desfontaines	(*Annual Beard Grass*)	Poaceae	Srinagar	1200	Used as fodder.
Polypogon monspeliensis (L.) Desfontaines	Annual *Beard Grass*	Poaceae	Kosi, Almora	1250	Used as fodder.

Contd...

Table 3.1–Contd...

Botanical Name	*Vernacular Name (English Name)*	*Family*	*Site/ Location*	*Altitude, m msl*	*Ethnobotanical Notes*
Pongamia pinnata (L.) Pierre	*Karanjua* (pongam Tree, Indian Beach)	Fabaceae	Kosi, Almora	1250	Seeds yield fatty oil; cultivated as roadside avenue.
Populus ciliata Wallich ex Royle	*Syan, Gar-peepal* (Poplar)	Salicaceae	Kosi, Almora	1200	Wood used for various purposes, mainly in paper industry, match-boxes and sports goods. Bark used as tonic, stimulant and blood purifier; leaves provide fodder.
Pouzolzia zeylanica (L.) J. Bennett and Brown		Urticaceae	Mahatgaon	1500	
Prunus armeniaca L.	*Chulu* (Apricot)	Rosaceae	Katarmal, Almora	1650	Fruits edible; seed oil edible and also used medicinally in fever and massaged in body pain; dried fruit pulp used in local brews; flowers useful in apiculture as bee-forage; leaves also used as fodder.
Prunus cera-soides D. Don	*Panyyan* (Himalayan Wild Cherry)	Rosaceae	Katarmal, Almora	1500	Branches for walking sticks; bark in psycho medicines; leaves as fodder; ripe fruits edible; juice from the bark applied on body swellings and contusions; flowers useful source of bee-forage; plant regarded as sacred, used in several rituals of local people.
Prunus cornuta (Wallich ex Royle) Steudel	*Jamna, Jamoni* (Himalayan Bird Cherry)	Rosaceae	Peethasain, Binsar	2400	Wood for construction purposes; leaves for fodder; ripe fruits edible; oil from the kernels applied on neuralgic pains; flowers useful in apiculture as bee-forage.

Contd...

Table 3.1–Contd...

Botanical Name	*Vernacular Name (English Name)*	*Family*	*Site/ Location*	*Altitude, m msl*	*Ethnobotanical Notes*
Pseudosorghum fasciculare (Roxb.) A. Camus		Poaceae	Dhikala	600	Grazed by animals.
Pterocarpus marsupium Roxb.	*Bija-Sal, Bija*	Fabaceae	Motadhang	400	Bark of the plant yields astringent gum known as 'Indian Kino'. Foliage provides good fodder. Timber valuable, used for various purposes.
Pueraria tuberose (Roxb. ex Willd.) DC.	*Siralu, Saral* (Indian Kutzu, Tuberous Honey Suckle)	Fabaceae	Srinagar	1200	Fibre obtained from the stem; leaves lopped for fodder, also provides fodder to elephants. Tubers made into various food preparations, supposed to provide strength; flowers an important source of bee-forage.
Pupalia lappacea (L.) Juss.		Amarantha-ceae	Abattmount, Lohaghat	1700	Fruits taken in cough and fever; externally applied on skin ailments.
Pyrus pashia Buch.-Ham. ex D. Don	*Mehal, Molu, Mole*	Rosaceaae	Byasghat, Bhrigukhal	1400	Leaves and twigs as fodder; ripe fruits edible and also used in digestive disorders; wood made into sticks; flowers useful in apiculture; checks soil erosion in landslide prone zones; important agroforestry tree.
Quercus floribunda Lindley ex Rehder	*Tilonj, Moru* (Green Oak)	Fagaceae	Binsar	2500	Wood used for household articles and as fuel; leaves provide fodder; fruits eaten by wild animals.
Quercus glauca Thunb.	*Harinj, Phaniyant* (Blue Japanese Oak)	Fagaceae	Champeshwar, Pauri	2000	Wood used as fuel, rarely for agricultural implements; leaves as fodder.

Contd...

Table 3.1–Contd...

Botanical Name	*Vernacular Name (English Name)*	*Family*	*Site/ Location*	*Altitude, m msl*	*Ethnobotanical Notes*
Quercus leucotrichophora A. Camus	*Banj* (White Oak)	Fagaceae	Tarag, Lohaghat	1650	Wood used for construction, plough, and bed-sticks, as well as fuel; leaves used as fodder; gum of the tree medicinally used for gonorrheal and digestive disorders; fruits eaten by monkeys and bears; decomposed leaves used as organic manure; an important tree of social forestry.
Quercus seme-carpifolia J.E. Smith	*Kharsu* (Brown Oak)	Fagaceae	Tarag, Lohaghat	2000	Wood used for construction and fuel; leaves as fodder; important tree of social forestry.
Rhamnus persica Boissiers	*Chirla* (Buck Thorn)	Rhamna-ceae	Sara	1200	Fruits edible; leaves browsed by goats.
Rhamnus purpureus Edgew.	*Gaunta, Luish*	Rhamna-ceae	Paboun	1500	Wood for agricultural implements; leaves lopped for fodder and manure; flowers useful source of bee-forage; bark extract with milk given in digestive disorders.
Rhus semialata Murray	*Deshmeel, Damphela*	Anacardiaceae	Tarag, Lohaghat	1650	Fruits edible and also used in folk medicines in colic and cholera; bark paste applied on body swellings and wounds.
Rhus wallichii Hook. f.	*Akhorie, Konkhee*	Anacardiaceae	Abattmount, Lohaghat	1700	Wood used for agricultural implements, especially to make tool handles; leaves lopped for cattle padding or as manure.
Rhynchosia minima (L.) DC.		Fabaceae	Srinagar	1200	As fodder; leaves regarded as abortifacient.

Contd...

Table 3.1–Contd...

Botanical Name	Vernacular Name (English Name)	Family	Site/ Location	Altitude, m msl	Ethnobotanical Notes
Rottboellia cochinchinensis (Lour.) W.D. Clayton	*Bhursali, Barsali* (Itch Grass)	Poaceae	Rishikesh, Kotdwar	400	Used for making mats; browsed by ovine.
Rubia cordifolia non L.	*Manjeet, Majethi* (Indian Madder)	Rubiaceae	Abattmount, Lohaghat	1700	Dye, commercially known as *Manjit*, extracted from the root and stem; roots medicinal, as tonic and astringent; stem used as an antidote to snake bite; flowers' extract in bacillary dysentery.
Rubia manjith Roxb. ex Fleming	*Majethi, Majeet, Lichkuru* (Indian Madder)	Rubiaceae	Bhanargaon	1400	Dye, commercially known as *Manjit*, extracted from the root and stem; roots medicinal, as tonic and astringent; stem used as an antidote to snake bite; flowers extract in bacillary dysentery.
Rubus ellipticus Smith	*Hisalu, Hinsar, Hinshoi* (Himalayan Yellow Raspberry)	Rosaceae	Patan, Lohaghat	1650	Fruits edible; root extract used in local beverages as intoxicating ingredient;flowers useful in apiculture as bee-forage; plant acts as soil binder; leaves rarely browsed by goats.
Rubus paniculatus Smith	*Kala-Hinsar, Kantula-Hinsar*	Rosaceae	Patan, Lohaghat	1650	Fruits edible; leaf paste externally applied on skin ailments.
Rumex dentatus L.	*Jangli-palak*	Polygonaceae	Matela, Almora	1200	Leaves scarcely used as vegetable.

Contd...

Table 3.1–Contd...

Botanical Name	*Vernacular Name (English Name)*	*Family*	*Site/ Location*	*Altitude, m msl*	*Ethnobotanical Notes*
Rumex hastatus D. Don	*Almoru, Kilmori*	Polygonaceae	Tarag, Lohaghat	1650	Sour leaves made into sauce; leaf extract applied on cuts and wounds to check bleeding and also believed to relieve from suffering of nettle sting.
Saccharum bengalensis Retz.	*Munja, Sarkanda, Sara*	Poaceae	Corbette Park	600	Yields fibre, used in cordages, ropes, mats and baskets, etc. Leaves in thatching, young plants as fodder.
Saccharum filifolium Steud		Poaceae	Khirsu, Pauri	1500	Use for fodder.
Saccharum longisetosum (Andersson ex Benth.) Narayanswami ex Bor	*Munj*	Poaceae	Sara, Pauri	1000	Leaves used as fodder.
Saccharum narenga (Nees ex Steudel) Hackel		Poaceae	Corbette National Park	600	Used as fodder.
Saccharum rufipilum Steudel		Poaceae	Patan, Lohaghat	1650	Culms used for thatching; young plants as fodder.
Saccharum spontaneum L.	*Munj, Kans, Kasa* (Thatch Grass)	Poaceae	Laldhang, Srinagar	800	Young plant used as fodder; leaves used in asthma and cholera.

Contd...

Table 3.1–Contd...

Botanical Name	*Vernacular Name (English Name)*	*Family*	*Site/ Location*	*Altitude, m msl*	*Ethnobotanical Notes*
Sacciolepsis indica (L.) A. Chase		Poaceae	Binsar	2000	Occasionally as fodder.
Salix acemophylla Boissier	*Bhains, Gad-bhains*	Salicaceae	Lansdowne	1400	Stem for making sticks; leaves as fodder.
Salix acutifolia Willd.	*Bhains, Bharroi* (Violet Willow)	Salicaceae	Binsar	2400	Twigs made into baskets and vessels; leaves and branches used as fodder.
Salix denticulata Anderson	*Bhotiana, Choti-Karvi*	Salicaceae	Adhwani	2600	Wood mainly used as fuel; leaves provide fodder; useful plant to check soil erosion.
Salix disperma Roxb. ex D. Don	*Bhaishara, Gad-bhains* (Willow)	Salicaceae	Adhwani	2600	Baskets are made from the branches; leaves used as fodder.
Salix tetrasperma Roxb.	*Gad-bhains, Jalmalya*	Salicaceae	Dugadda	1000	Wood for agricultural implements and basketry; leaves as fodder; decoction of bark used to relieve fever; flower as a source of bee-forage.
Salvia hians Royle ex Benth	*Ghanya, Goani, Ganeeth*	Lamiaceae	Chaubatiya, Almora	2250	Leaf juice rubbed over the body swelling, arthritic pain and eczema.

Contd...

Table 3.1–Contd...

Botanical Name	*Vernacular Name (English Name)*	*Family*	*Site/ Location*	*Altitude, m msl*	*Ethnobotanical Notes*
Salvia leucantha Cav.	*Ganya, Ghaniya*	Lamiaceae	Chauri, Lohaghat	1650	Leaf juice rubbed over the body swelling, arthritic pain and eczema; leaf infusion given in colic and diarrhoea; flower paste used in cold and cough; root extract used in cough and cold; powder of seeds useful in gonorrhoea and menorrhoea; flowers as a source of bee-forage.
Saurauia nepaulensis DC.	*Gugna, Ratendu, Gogina*	Saurauiaceae	Lansdowne	1600	Ripe fruits edible; wood for construction and agricultural implements; leaves lopped for fodder.
Schleichera oleosa (Lour.) Oken	*Kusum* (Ceylon Oak, Honey Tree, Lac Tree)	Sapindaceae	Neelkanth	1200	Sub acid pulp (aril) of the fruit edible; wood durable, used for construction purposes; oil obtained from the seeds used for burning, hair, and medicinal purposes; leaves as fodder; useful source of bee-forage; an important tree of social forestry.
Scurrula cordifolia (Wallich) G. Don	*Banu*	Lorantha-ceae	Chamkotkhal	400	Occasional fodder.
Scurrula elata (Edgew.) Danser	*Banu, Bandu*	Lorantha-ceae	Binsar	2400	Occasional fodder.
Sesbania sesban (L.) Merrill	*Jainti, Rawasan*	Fabaceae	Dugadda	1000	Used as hedge plant due to fast growth; stem yields fibre used for cordages and ropes; foliage provides excellent fodder; also made into manure.
Setaria barbata (Lam.) Kunth		Poaceae	Corbette National Park	600	Widely used as fodder.

Contd...

Table 3.1–Contd...

Botanical Name	*Vernacular Name (English Name)*	*Family*	*Site/ Location*	*Altitude, m msl*	*Ethnobotanical Notes*
Setaria glauca (L.) P. Beauv.	*Ban-Kauni* (Bottle Grass)	Poaceae	Chanmari, Lohaghat	1650	Fodder for livestock; grains occasionally eaten, often used in local beverages.
Setaria homonyma (Steudel) Choivenda		Poaceae	Chauri, Lohaghat	1650	Fodder for cattle.
Setaria intermedia Roemer and Schultes		Poaceae	Matela, Almora	1200	Widely used as fodder.
Setaria pumila (Poiret) Roemer		Poaceae	Katarmal, Almora	1650	Used as fodder.
Setaria verticillata (L.) P. Beauv.	*Latkaunya, Laptuna* (Rough Bristle Grass)	Poaceae	Sara, Srinagar	1200	Widely used as fodder. Mature grains provide food for animals. Pudding of grains taken in constipation.
Setaria viridis (L.) P. Beauv.	*Birali-ghas* (Green Bristle-Grass)	Poaceae	Daira-Binsar	2500	Grains edible; hay and green grass as fodder.
Shuteria involucrate (Wallich) Wight and Arn.		Fabaceae	Kumbhichaur	800	Used as fodder.
Shuteria vestita Wight and Arn.		Fabaceae	Rudraprayag	1500	Plant used as fodder.

Contd...

Table 3.1–Contd...

Botanical Name	*Vernacular Name (English Name)*	*Family*	*Site/ Location*	*Altitude, m msl*	*Ethnobotanical Notes*
Sinarundinaria anceps (Mitf.) Chao and Renvoize	*Saruru, Ringal*	Poaceae	Chaubatiya, Almora	2250	Widely used for wicker-works; cottage, mats and baskets; leaves as fodder.
Sinarundinaria falcate (Nees) Chao and Renoize	*Gad-ringal, Ringal*	Poaceae	Chaubatiya, Almora	2250	Widely used for wicker works; cottage, mats and baskets; leaves as fodder.
Smilax aspera L.	*Kukurdara, Kumardar* (China Root)	Smilaca-ceae	Chanmari, Lohaghat	1650	Occasionally young leaves are cooked. Roots are diuretic and diaphoretic; root paste with mustard oil massaged on rheumatic-arthritis; flowers source of bee-forage.
Smilax glaucophylla Klotzsch	*Kukurdara, Sarya*	Smilaca-ceae	Bharsar	2200	Root decoction given in flatulence and aphthae; supposed to be useful in fever; plant lopped for fodder; occasionally young shoots used as vegetable.
Smithia ciliate Royle		Fabaceae	Dadamandi	1000	Plant provides excellent fodder.
Solanum nigrum L.	*Makoi, Kirmoi, Palpotan* (Black Nightshade)	Solana-ceae	Chauri, Lohaghat	1650	Fruits edible; plant extract used in liver disorders, piles and dysentery; fruits useful in diarrhoea, fever and eye ailments; occasionally roots added to local beverages.
Sonchus brachyotus DC.	*Karatu* (Corn Sow Thistle)	Asteraceae	Abattmount, Lohaghat	1700	Roots used in the folk medicine against cough and bronchitis; young shoots and leaves as vegetable in famine.

Contd...

Table 3.1–Contd...

Botanical Name	*Vernacular Name (English Name)*	*Family*	*Site/ Location*	*Altitude, m msl*	*Ethnobotanical Notes*
Sorghum halepense (L.) Persoon	*Chinna* (Johnson Grass)	Poaceae	Rasulpur, Laldhang	600	Provides good fodder.
Sorghum nitidum (Vahl) Persoon		Poaceae	Kotdwar	300	Used as fodder, also made into brooms.
Spatholobus parviflorus (Roxb. ex DC.) Kuntze	*Malini, Maula, Bando*	Fabaceae	Sara	600	Stem provides fibre, like *Bauhinia vahlii*. Leaves preferred for fodder.
Spermadictyon sauveolens Roxb.	*Padera, Padar*	Rubiaceae	Ghingharikhal, Almora	2000	Root powder as extract given in diarrhoea and cholera; leaves as fodder and also used as insecticide for the stored grains; flowers as an important source of bee-forage.
Sporobolus diander (Retz.) P. Beauv.	*Sitya*	Poaceae	Chanmari, Lohaghat	1500	Crushed leaves applied on burns and pimples; supposed to be good soil-binder; also provides fodder.
Sporobolus fertilis (Steudel) Clayton		Poaceae	Srinagar	1200	Provides fodder.
Sporobolus piliferus (Trinius) Kunth		Poaceae	Ghingharikhal, Almora	1950	Fodder for cattle.

Contd...

Table 3.1–Contd...

Botanical Name	*Vernacular Name (English Name)*	*Family*	*Site/ Location*	*Altitude, m msl*	*Ethnobotanical Notes*
Sporobolus spicatus (Vahl) Kunth		Poaceae	Khirsu	1800	Occassionally used for fodder.
Stellaria media (L.) Villars	*Badyalu* (Chick Weed)	Caryophylla-ceae	Ghingharikhal, Almora	2000	Plant used as green vegetable as well as fodder; plant paste externally applied in burns, boils and wounds.
Stipa roylei (Nees) Mez		Poaceae	Nainidanda, Dhumakot	2300	Fodder for cattle.
Stipa sibirica (L.) Lam.		Poaceae	Pauri-Khirsu	2300	Fodder for cattle, although believed to be poisonous.
Streblus asper Lour.	*Dahia, Siora* (Siamese Rough-Bush)	Moraceae	Corbette Park	600	Branches lopped for fodder; fruits edible; wood used for household articles; leaf paste is useful in skin ailments.
Strobilanthes atropurpureus Nees	*Kangadai, Totna*	Acanthaceae	Abattmount, Lohaghat	1700	Young shoots cooked as vegetable in scarcity.
Swertia angustifolia Buch.-Ham. ex D. Don	*Chirata*	Geraniaceae	Abattmount, Lohaghat	1700	Used in Ayurvedic medicine 'Chiretta', which is used in blood diseases and as febrifuge.
Swida macrophylla (Wallich) Sojak	*Khagsi* (Himalayan Dog Wood)	Cornaceae	Chaubatiya, Almora	2250	Wood for agricultural implements; leaves lopped for fodder; flowers useful inapiculture as bee-forage; fruits edible; an important plant of social forestry.

Contd...

Table 3.1–Contd...

Botanical Name	*Vernacular Name (English Name)*	*Family*	*Site/ Location*	*Altitude, m msl*	*Ethnobotanical Notes*
Swida oblonga (Wallich) Sojak	*Katkanai*	Cornaceae	Almora	1700	Wood useful for agricultural implements; leaves provide fodder.
Symplocos paniculata (Thumb.) Miq.	*Lodh* (Sapphire Berry)	Symploca-ceae	Binsar	2400	Flowers sucked for nectar by bees; yellow dye obtained from the bark. Leaves lopped for fodder; bark used in folk medicine to check abortion; useful in apiculture as bee-forage.
Symplocos racemosa Roxb.	*Lodhra, Lodh*	Symploca-ceae	Baganikhal	1000	Bark and leaves yield yellow dye; leaves as fodder; bark is also used in indigenous medicines; good source of bee-forage.
Symplocos ramo-sissima Wallich ex G. Don	*Lodh*	Symploca-ceae	Adhwani, Binsar	2500	Leaves used as fodder; flowers visited by bees for nectar and pollen.
Taraxacum officinale Weber	*Kanphuliya, Karatu, Dudhee* (Common Dandleon, Bitterwort)	Asteraceae	Patan, Lohaghat	1650	Root extract used in the treatment of migraines, hepatitis and headache; young shoots and leaves, thoroughly boiled and cooked as vegetable.
Tephrosia candida (Roxb.) DC.	*Ban-tor* (White Tephrosia)	Fabaceae	Srinagar	1200	Leaves used as fodder; bark used to intoxicate fishes.
Tephrosia pumila (Lam.) Persoon		Fabaceae	Kyar	500	Occasionally used as fodder.

Contd...

Table 3.1–Contd...

Botanical Name	*Vernacular Name (English Name)*	*Family*	*Site/ Location*	*Altitude, m msl*	*Ethnobotanical Notes*
Tephrosia purpurea (L.) Persoon	*Pal, Sar-punkha*	Fabaceae	Kateburgh	800	The plant extract used in hepatic disorders; decoction of pods as vermifuge; good soil-binder as well as fodder.
Terminalia alata Heyne ex Roth	*Asin*	Combreta-ceae	Mahatgaon, Almora	1500	Wood used for construction, house building and charcoal; leaves as fodder as well as host of tasar silk worms; bark used in liver troubles; flowers as source of bee-forage in apiculture; an important tree of social forestry.
Terminalia chebula Retz.	*Haira, Har, Harara*	Combreta-ceae	Katarmal, Almora	1600	Wood used for various purposes; leaves as fodder; kernel of fruit edible; the 'black myrobalan' is the dried fruit of the plant, used as one of the constituents of 'triphala'; much priced for medicines; the galls of stem or leaves yield yellow dye; flowers useful in apiculture as bee-forage; an important plant of social forestry.
Tetrastigma lanceolarium (Roxb.) Planchon		Vitaceae	Matela, Almora	1200	Leaves provide favourite fodder.
Thalictrum foliolosum DC.	*Mamiri, Kirmuli* (Indian Meadow Rue)	Ranuncula-ceae	Tarag, Lohaghat	1650	Roots used in ophthalmia and also in colic fever.
Thamnocalamus falconeri Hook. f. ex Munro	*Devaringal*	Poaceae	Ghingharikhal, Almora	2200	Used in thatching of roofs, in making mats, baskets, etc; provides good fodder.

Contd...

Table 3.1–Contd...

Botanical Name	*Vernacular Name (English Name)*	*Family*	*Site/ Location*	*Altitude, m msl*	*Ethnobotanical Notes*
Thamnocalamus spathiflora (Trinius) Munro	*Tham-ringal, Ningal*	Poaceae	Chaubatiya, Almora	2300	Used in thatching of roofs, in making mats, baskets, etc; provides good fodder.
Themeda anathera (Nees ex Steudel) Hackel	*Ghatira*	Poaceae	Patan, Lohaghat	1650	Dried plants used as fodder, young one believed to be poisonous.
Themeda arundenacea (Roxb.) Ridley	*Kalpura, Kapur-Ghas, kangua* (Ulla Grass)	Poaceae	Kosi, Almora	1200	Root extract given in fever; used as fodder; also used in thatching.
Themeda triandra Forsk		Poaceae	Dhumakot-Binsar	2000	Fodder for cattle.
Themeda villosa (Poiret) A. Camus		Poaceae	Corbette National Park	600	Fodder for animals.
Thymus biflorus Buch.-Ham. ex D.Don	*Gorakhopan, Ban-ajwain* (Indian Wild Thyme)	Lamiaceae	Abattmount, Lohaghat	1700	Flavour of crushed leaves inhaled in cold and sinusitis; extract of leaves with milk given in gastroenteritis.
Thysanolaena maxima (Roxb.) Kuntze	*Naktura, Nastura* (Bouquet Grass, Broom Grass)	Poaceae	Sara, Srinagar	1200	Fodder and broom grass; grains abortifacient.

Contd...

Table 3.1–Contd...

Botanical Name	*Vernacular Name (English Name)*	*Family*	*Site/ Location*	*Altitude, m msl*	*Ethnobotanical Notes*
Tinospora sinensis (Lour.) Merrill	*Giloi, Gilai* (Tinospora)	Menispermaceae	Katarmal, Almora	1650	Juice of stem and leaves is used in general debility and urinary troubles; leaves as fodder; extract regarded as an emetic; also useful in leprosy; plant decoction given to relieve malarial fever.
Toona ciliata Roemer	*Tun* (Cederela Tree)	Meliaceae	Tarag, Lohaghat	1650	Wood priced for construction purposes, furniture and other articles; sulphur dye is obtained from the flowers; useful source of bee-forage and apiculture; important tree of social forestry.
Toona serrata (Royle) M. Roemer	*Daul, Pahari-tun* (Indian Mahogony, Hill Toon)	Meliaceae	Bironkhal, Champeshwar	1800	Timber-yielding plant, leaves lopped for fodder; bark paste applied on ulcers and boils; the saw-dust of the plant after boiling and decanting regarded useful in asthma; useful tree of social forestry.
Torenia cordifolia Roxb.		Scrophulariaceae	Abattmount, Lohaghat	1700	Occasionally as fodder.
Tragus roxburghii Panigrahi		Poaceae	Digoli	600	As fodder.
Trema orientalis (L.) Blume	*Jivan* (Nettle Tree)	Ulmaceae	Matela, Almora	1200	Bark yields strong fibre, leaves fodder; fruits edible; wood used for charcoal.
Trema politoria (Planchon) Blume	*Koeli, Khagshi*	Ulmaceae	Dugadda, Satpuli	1200	Bark yields strong fibre; wood for fuel and charcoal; leaves considered as a good fodder; fruits edible, sometimes made into jam and also used to check vomiting.

Contd...

Table 3.1–Contd...

Botanical Name	*Vernacular Name (English Name)*	*Family*	*Site/ Location*	*Altitude, m msl*	*Ethnobotanical Notes*
Trichosanthes palmata Roxb.	*Indrian, Lal Indriyan*	Cucurbitaceae	Abattmount, Lohaghat	1700	Roots and fruits medicinal, commonly used to relieve bronchitis, asthma, diabetes; root juice used as an emetic; seed paste applied externally on hoops and mouth disease.
Trifolium alexandrinum L.	*Berseem* (Egyptian Clover)	Fabaceae	Matela, Almora	1200	As an important source of bee-forage and as fodder.
Trifolium repens L.	*Tipatiya* (White Clover, Dutch Clover)	Fabaceae	Khetikhan, Lohaghat	1650	Leaf paste used as an astringent; plant provides good fodder.
Tripogon filiformis Nees ex Steudel		Poaceae	Patan, Lohaghat	1650	Used for fodder.
Ulmus wallichiana Planchon	*Chamarmora, Mairu*	Ulmaceae	Adhwani	2500	Bark yields strong fibre; leaves yield fodder. Plaster used for fractured bones is made by boiling the bark in water.
Urena lobata L.	*Chatkura, Unga*	Malvaceae	Patan, Lohaghat	1650	Stem yields a coarse fibre; flowers expectorant; root paste applied on body pain and in rheumatism.
Urochloa panicoides P. Beauv.		Poaceae	Digoli	800	Used as fodder.

Contd...

Table 3.1–Contd...

Botanical Name	*Vernacular Name (English Name)*	*Family*	*Site/ Location*	*Altitude, m msl*	*Ethnobotanical Notes*
Urtica dioica L.	*Kandali, Bicchhu-ghas* (stinging Nettle)	Urticaceae	Chaubatiya, Almora	2250	Stem yields strong-shining fibre, used for sacs, ropes and mats; young branches and leaves used as delicious pot herb; seed oil edible as well as medicinal in sciatica, rheumatism and several skin ailments; hair-wash from leaf extract believed to avoid baldness.
Urtica parviflora Roxb.	*Bicchu-ghas*	Urticaceae	Chauri, Lohaghat	1650	Stem yields strong-shining fibre, used for sacs, ropes and mats; young branches and leaves used as delicious pot herb; seed oil edible as well as medicinal in sciatica, rheumatism and several skin ailments; hair-wash from leaf extract believed to avoid baldness.
Vetiveria zizanioides (L.) Nash	*Khas, Veeran-mool* (Khus-Khus, Vetiver)	Poaceae	Kalaghati-Bhabar	600	Root aromatic, used as anthelmintic and in burns and boils; hanging mats are made to act as cooler; fodder plant also.
Viburnum coriaceum Blume	*Lampatya, Rashina, Kala Titmaliya*	Caprifolia-ceae	Abattmount, Lohaghat	1700	Fuel as well as fodder; flowers useful source of bee-forage.
Vicia sativa L.	*Kurphali* (Common Vetch)	Fabaceae	Chakisain	1200	As fodder.
Vicia tenera Graham ex Benth.	*Kura*	Fabaceae	Matiyali, Thalisain	1800	Used as fodder.

Contd...

Table 3.1–Contd...

Botanical Name	*Vernacular Name (English Name)*	*Family*	*Site/ Location*	*Altitude, m msl*	*Ethnobotanical Notes*
Vigna aconitifolia (Jacquin) Marechal	*Matki, Moth* (Moth Bean, Mat Bean, Aconite Bean)	Fabaceae	Mandul-Paukhal	1200	Young pods and seeds are edible; whole plant used as fodder.
Vigna sublobata (Roxb.) Babu and Sharma		Fabaceae	Matiyali	1600	Pods and seeds edible; whole plant used as fodder.
Vigna trilobata (L.) Verdc.	*Moth, Maugam*	Fabaceae	Chelusain	800	Whole plant used as fodder.
Vigna vexillata (L.) A. Richard	*Machali*	Fabaceae	Srinagar	1200	Fusiform roots edible; plant as fodder.
Viola biflora L.	*Vanafsa* (Yellow-Violet)	Violaceae	Dudhatoli	2700	The whole plant either in the form of extract or powder taken as diaphoretic, useful in skin and blood diseases; flowers and leaves boiled with tea, supposed to be good for fever and cough. Plant is browsed by sheep and goats. taken as diaphoretic,
Vitis flexuosa Thunb.	*Panlaguli, Rampauren*	Vitaceae	Chaubatiya, Almora	2200	Leaves as fodder; fruits edible; flowers useful source of bee-forage.
Wrightia tomentosa (Roxb.) Roemer and Schultes	*Darula, Daira, Dudhi*	Apocyna-ceae	Mahatgaon, Almora	1450	Bark yields a yellow dye, also used as an antidote to snake bite; wood used for preparing various carved articles; fruits and seeds sometimes edible; latex useful in wounds and cuts.

Contd...

Table 3.1–Contd...

Botanical Name	Vernacular Name (English Name)	Family	Site/ Location	Altitude, m msl	Ethnobotanical Notes
Ziziphus mauritiana Lam.	*Ber* (Jujube)	Rhamnaceae	Warm valleys, Almora	600	Fruits edible, wood for construction and agricultural implements; flower useful in apiculture as source of bee-forage; important plant of social forestry.
Ziziphus nummularia (Burm. f.) Wight and Arn.	*Jharberi* (Wild Jujube)	Rhamnaceae	Kotdwar	400	Fruits edible; fodder from leaves.
Ziziphus . xylopyrus (Retz.) Willd	*Kath-ber*	Rhamnaceae	Kotdwar	400	Fruits edible; wood occasionally used as fuel; flowers as source of bee-forage; plants check soil erosion.

Note: Ethnobotanical values are based on Gaur (1999). Virtually all plants have some fodder value; these are grazed or browsed by domestic and wild animals, or their leaves are lopped for feeding to livestock at stall. Some plants are browsed only by certain species of animals, not by all.

variety of grass species. Several types of grass covers occur in the areas. Dabadghao and Shankarnarayan (1973) described five major types of grasslands in India out of which two types, *viz.*, *Themeda-Arundinella* (500-2500m) and temperate alpine zones (above 2500 m), were recorded in the Himalaya. In the Western Himalaya, *Cenchrus ciliaris, Cynodon dactylon, Bothriochola pertusa* grass cover is dominant up to 500m elevation. The zone between 500-2500m is dominated by *Chrysopogon fulvus, Themeda anathera, Arundinella nepalensis* and *Heteropogon contortus*. Other common grasses in this zone are *Eulaliopsis binata, Chryspogon gryllus, Muhlenbergia duthieana, Cymbopogon stracheyi. Bromus ramosus, Dictylis glomerata* and *Oryzopsis acquiglummis*. Regions above 3300m were represented by *C. gryllus, Poa annua* and *Agrostis canina*. Agarwal (1959) reported 4 grasslands types in the Chakrata forest division of Dehradun district. Gupta (1974) has described 9 types of grass covers from the Central Himalaya (Garhwal Himalaya). Numata (1983) and Tsuchida (1983) reported that *Dactyloctenium aegypticum, Chrysopogon aciculatus, Cynodon dactylon, Imperata cylindrica, Paspalum scrobiculatum, Cyperus rotundus, Arundinella, Dichanthium, Setaria pallidafusca, Arthraxon quartinianus, Eragrostis, Panicum* and *Paspalidium* species of grasses were common on sub-tropical and temperate zones of the eastern Nepal. Also in road sides on steep slopes, a few grass communities occur; however, grass species were found in all kinds of vegetations. About 250 species of Western, Central and Eastern Himalaya were presented by Patil and Pathak (1980). When the grasslands are subjected to grazing on burning, a variety of grassland types develop which vary in successional status (Gupta and Nanda 1970, Dabadghao and Shankarnarayan 1973, Gupta 1974, Patil and Pathak 1980, Numata 1983, Tsuchida 1983, Sundriyal *et al.*, 1987, Singh and Gupta 1993). The diversity of fodder plants, as enlisted after this exploration, is a proportion of the enormous biodiversity occurring in the parts of Himalaya, *e.g.*, in the Garhwal district of Uttarakhand, a profile of which has been beautifully presented by Gaur (1999). He has enumerated as many as 2150 species belonging to 1032 genera and 189 families in the Garhwal district of Uttarakhand Himalaya. His work underlines the very high degree of biodiversity in mountain habitats. Much of the proportion of this biodiversity is found in the rangeland ecosystems and comprises fodder for livestock. Gaur (2004), Gaur and Semwal (1983), Gaur and Tiwari (1987), and Gaur *et al.* (1983, 1984, 1985, 1988, 1992) have also presented the diverse scenarios and various uses of the vegetation prospering in Kumaun and Garhwal Himalaya.

A few wild plants are found to grow on the vertical stairs of the fields or on margins, playing an important ecological role in maintaining the soil fertility. These plants, according to Gaur (1999), form an intermittent strip in between two crop fields, balancing effects of cultigens. These vegetationally rich zones provide essential elements by their death and decay and subsequently leached to add soil fertility of immediate field, besides maintaining thermodynamics, as well as humidity of the croplands. The plants grown in the border area of the fields evidently check soil and water erosion (Gaur 1999). Some of the important plants that have been found are *Artemisia nilagirica, Arthraxon lanceolatus, Arthraxon prionodes, Arundinella bengalensis, Arundinella nepalensis, Barleria cristata, Bidens biternata, Bidens pilosa, Boehmeria platyphylla, Bupleurum hamiltonii, Capillipedium parviflorum, Crotalaria albida, Cyperus corymbosus, Cyperus monocephalus, Cyperus nutans, Cyperus rotundus, Desmodium laxiflorum, Desmodium microphyllum, Digitaria adscendens, Erigeron bonariensis, Galium aparine, Galium elegans, Girardinia diversifolia, Imperata cylindrica, Leptodermis lanceolata, Oplismenus compositus, Panicum antidotale, Panicum miliaceum, Panicum psilopodium, Pennisetum orientale, Polygonum capitatum, Polygonum hydropiper, Pouzolzia zeylanica, Rubia manjith, Rubus ellipticus, Rumex dentatus, Rumex hastatus, Saccharum bengalensis, Saccharum spontaneum, Salvia hians, Spermadictyon sauveolens, Thalictrum foliolosum, Themeda anathera, Urtica dioica,* etc.

It is revealed that livestock graze selectively on some species and they prefer nutritious forbs (Gupta 1986, Sundriyal and Joshi 1991), and thus unpalatable species proliferate in the grassland. Yadava and Singh (1977) argued that succession in grasslands often intimately associated with grazing pressure. It can be concluded that multitude of factors, *i.e.*, altitude, slope, aspect, canopy cover, soil types, biotic pressure and successional status determine the grassland composition in any area, and extent of these factors vary from place to place (Patil and Pathak 1980, Sexena and Singh 1980, Singh and Sexena 1978, Melkania and Tandon 1983, Gupta 1986, Melkania and Singh 1989).

Species composition of the grassland at highest stage of progression, naturally, includes most productive and nutritious species. However, with increasing biotic pressure, vegetation cover is reduced significantly. When the root reserves of energy are exhausted as a result of frequent defoliation, the grasses cease to regenerate and the rangelands suffer from eco-crises.

Fodder Species in High Altitude and Alpine Zones

Alpine zones (3000-5000m msl) usually do not represent tree species, and the herbage covering the major parts of the meadows and slopes are often perennial and in the form of tussocks, widely found on slopes and meadows (3000-4000m msl). *Danthonia cachemyriana, D. exilis, Festuca kashmiriana, F. polycolea* and *Stipa roylei* occupy the major area in the dense tufted tussocks, widely found on slopes and meadows (3000-4000 m msl). Above 4000m, tussocks grasses are restricted in distribution on meadows, moraines, crevices, streams or glaciers. The common grasses of meadows and moraines are *Briza media, Bromus ramosus, B. tectorum, Calamagrostis holciformis, C. pseudophragmites* var. *tartarice, Festuca parvigluma, F. valesiaca, F. rubra* var. *villosa, Phleum alpinum, Poa elonare, P. hirtiglumis, P. pagophila, P. supina, P. stapfiana, Puccinelia thomsonii, Trisetum spicatum,* etc. Grasses found in the glacier deposition are *Calamagrostis lahulensis, Poa bulbosa, P. kanaii, P. mustangensis, Trisetum spicatum,* etc. (Nautiyal 1996).

Some of the grasses show wide distribution from submontane to montane to alpine zones in different habitats. Species of *Apluda, Arundinella, Bothriochloa, Chrysopogon, Cymbopogon, Cynodon, Digitaria, Microstegium,* and *Poa* represent wider altitudinal distribution from submontane to montane zones. *Poa annua* is found to occur from plains to alpine zones (4000m msl). *Agrostis vinealis, Bromus japonicus, Calamagrostis emodensis, C. lahulensis, Danthonia cachemyriana, Dactylis glomerata, Phleum alpinum, Poa himalaya, Stipa roylei,* etc. are distributed from montane to alpine zones (Nautiyal 1996).

The Poaceae comprises the most valuable family of flowering plants, as they provide staple food in the form of cereals to human beings and forage to domestic and wild animals, including several herbivores of lower groups. Besides, grasses provide sugar, medicines, beverages, edible or essential oils, paper-pulp, fibres for large number of articles, etc. (Nautiyal 1996, Gaur 1999).

Family Poaceae frequently forms dominant vegetation in various types of habitats, particularly in high altitudes and alpine zones of the Himalaya. Some of the species of this family occurring in different habitats are listed in Table 3.2.

Table 3.2: Plants of Family Poaceae Occurring in Different Habitats in Uttarakhand

Name of the Plant	*Location*	*Altitude, m msl*
Poa nephelophila	Uttarkashi	3300
Poa kanaii	Uttarkashi	4000
Poa mustangensis	Uttarkashi	4000
Agrostis gigantea	Pithoragarh	3000
Bromus mollis	Pithoragarh	3000-3500
Calamagrostis epigeois	Uttarkashi	2500-3000
Danthonia exilis	Uttarkashi	3900-4000
Elymus dahuricus	Pithoragarh	3500-3900
Festuca rubra var. villosa	Pithoragarh	3900-4200
Oryzopsis brachyclada	Uttarkashi	3000
Poa aitchisonii	Chamoli	3000
Poa bacteriana	Uttarkashi	4000
Poa palustris	Chamoli	3300
Poa setulosa	Chamoli	3000-3500
Puccinellia thomsonii	Uttarkashi	3500
Agrostis filipes	Uttarkashi	3300
Arostis hookeriana	Uttarkashi	3900
Danthonia communis	Pithoragarh	3500
Festuca polycolea var. *brevis*	Chamoli	3000
F. rubra var. *clarkei*	Pauri	2300-2500
Helictotrichon parviflorum	Chamoli	3000
Phragmites karka var. cincta	Tehri	2400
Poa eleanorae	Chamoli	3000-3300
Poa hirtiglimis	Uttarkashi	3500-3900
Poa khasiana	Uttarkashi	3900

Grasses cover some 20 per cent of the global area. Role of grasses as soil binders, and indicator value of soil conditions or ecological

Plate 3.1: A Mixed Type Rangeland as a Rich Repository of Fodders

Plate 3.2: *Grewia optiva* in an Agroforestry System

Plate 3.3: ***Machilus duthiei*** **(Kaul) in a Well Managed Rangeland**

Plate 3.4: ***Ficus clavata*** **and Grasses in a Rangeland**

Plate 3.5: *Quercus leucotrichophora* (Oak): A Symbol of Ecological and Socio-economic Prosperity and Posterity of Life in the Himalayan Mountains

Plate 3.6: A Conifer Type Rangeland

Plate 3.7: Grass-dominated Rangeland

Plate 3.8: Rangeland on Abandoned Cultivated Area

Plate 3.9: Mixed Forest Ecosystem as a Backbone for Mountain Agriculture and Livelihoods

Plate 3.10: Grass Protected in a Rangeland for Rich Harvest followed by Hay Making

Plate 3.11: A Mixed Type Rangeland is Always More Stable and More Promising

Plate 3.12: A Rangeland as a Natural Repository of Biodiversity and Rich Repository of Species of the Poaceae Family

Plate 3.13: Oak Type Forest Ecosystem as a Cornucopia of Socio-economic and Ecological Attributes

Plate 3.14: Degraded Rangelands are Posing Threat to Fodder Supplies

All photos: Babita Bohra

Plate 3.15: Rangelands are the Basis for a Livestock-based Livelihood in the Mountains (Photo: Gita Singh)

status of particular locality is immensely recognized. Nevertheless, grasslands/grazing lands in the Himalaya are seldom provided due attention in development intervention.

Ethnobotanical Values of Fodder Species

Apart from fodder value, many of the plants also provide edible fruit, fuel wood, fibre, flowers, buds, vegetables, seeds, dye, bee-forage, etc. Some species have unique ecological role, such as specific soil-binding properties, water conservation, pollination, etc. Some plants are of crucial ethnobotanical importance and are brought into various uses, such as in curing of certain diseases, amelioration of health, religious rituals, cultural rites, etc. Ethnobotanical notes are also given against most of the fodder species occurring in the mountain rangelands (Table 3.1).

Different parts (such as root, shoot, leaves, flowers, and seeds) of a species are used through certain physical and chemical processes in the curing of prevailing diseases, such as skin and eye problems, diarrhoea, dysentery, digestive problems, menstrual disorders, gonorrhoea, dysmenorrhoea, measles, respiratory problems, bronchitis, fever, malaria, dropsy, piles, diabetes, jaundice, suppressed urination, urinogenital problems, delivery-related complexities, etc. Many fodder species are of ethno-veterinary uses and people variously apply them in livestock health management. Certain species are regarded as sacred and are used in various cultural and religious rites. Local people in Uttarakhand mountains, as also in other parts of the Himalaya, are well equipped with the knowledge relating to the various uses of plants.

This traditional knowledge is millennia old and each generation has been passing it on to the next generation. Management (conservation and utilization) of fodder resources thus exhibits one of the unique examples of ethnobotanical applications in resource management by range-dependent mountain communities.

Chapter 4
Wildlife in Range Ecosystems

Diverse Himalayan habitats comprise different composition of vegetation or phyto-community. The vegetation types give rise to specific set of microclimatic conditions in which there is a definite pattern of energy flows, *i.e.*, food chain, and in which specific species of wild animals thrive. Himalayan biota have been under five different biogeographic influences (Khoshoo 1995), *viz.*, Palaeoarctic, Mediterranean, Sino-Japanese, Indo-Malayan, and Peninsular Indian. Even then there is considerable endemism and uniqueness associated with Himalayan biodiversity. According to an old study by Chatterjee (1939), there are 6,850 flowering plants endemic to India, out of which 3,165 are endemic to the Himalaya. The Himalayan region, thus, comprises very high percentage (about 50 per cent) of India's endemic flora. Again, there is very high degree of endemism in wildlife that organizes in harmony with specific habitats. Major wild fauna inhabiting diverse ecosystems in the Himalaya includes elephant, leopard, goral, clouded leopard, rhinoceros, Manipur deer, wild buffalo, pheasant birds and reptiles.

Fauna of the Indian Himalaya

Himalayan fauna has not been assessed adequately. Amongst the invertebrates, it is only butterflies that have been accounted for to an appreciable extent. Other insects have not been brought under thorough exploration. Reptiles, fish, and amphibians have been covered only partially. Birds and mammals, on the other hand, have been described to a great degree.

Avifauna of the Himalaya is a mixture of Palaeoarctic and Indo-Chinese elements. Out of 2,100 species of birds, nearly 300 are winter visitors from the Palaeoarctic Region north of the Himalayan barrier (Ali 1981). Ducks, geese, cranes, and wading and shore birds which become very prominent during winter on many water bodies in India, are the main amongst the 'long-distance extra-limital migrants' (Khoshoo 1995). Siberian-Central Asian route and Tsangpo or Brahmaputra River route are the important routes of these birds.

The most distinctive family of birds in the Himalaya is Phasianidae. This includes birds like pheasants, jungle fowl, partridges, and quails. These together are known as game birds. The Himalayan pheasants–that include some of the most spectacular and gorgeously plumaged birds in the world–have their centre of primary distribution in the Indochina sub-region. The prominent pheasants are blood pheasant (*Ithanginis cruentus*), horned pheasant (*Tragopan satyra*), monal (*Lophophorus*), Kalij pheasant (*Lophura teucomelana*), Koklas (*Pucrasia macrolopha*), and chir pheasant (*Catreus wellichii*). Their distinctiveness attracts aviculturists, zoos, poachers, smugglers, and private collectors to tame them. In fact, this group of birds is the most sought after one and now is at the verge of extinction (Khoshoo 1995).

The Greater Himalayan areas provide a harsh and somewhat inhospitable habitat for many animals. Many big animals are not adapted to these high altitudes. But there are some wild animals that flourish in this habitat. Animals of goat family and some species of antelope family–especially the *chiru* (goat antelope, *Pantholops hodgsoni*) and musk deer (*Moschus moschiferus*)–found in this region are now highly endangered on account of habitat loss and overexploitation (Ranjitsinh 1981).

The brown bear (*Ursus arctos isabellinus*)–an omnivore–is the largest animal in the Greater Himalaya. The black bear (*Selenoroctos thibetanus*) has wide range of habitat from the western to the eastern

Himalaya. Main canine in the Greater Himalayan reaches is the snow leopard (*Panthera uncia*). The red panda (*Ailurus fulgens*)–totally vegetarian and adapted to a very narrow range of habitat–is the counterpart of the Chinese giant panda but is smaller in size (Khoshoo 1985).

Fauna of the Uttarakhand Himalaya

Diverse Himalayan habitats comprise different composition of vegetation or phyto-community. The vegetation types give rise to specific set of microclimatic conditions in which there is a definite pattern of energy flows, *i.e.*, food chain, and in which specific species of wild animals thrive. In the Central Himalaya–the State of Uttarakhand–the following vegetation types have been recognized (after Singh and Singh 1987, Ram *et al.*, 1993):

1. ***Submontane seasonal broadleaf forests*** are found at the lower altitudes from foothills to 1000 m. Sal (*Shorea robusta*) is the dominant and the most extensively distributed species. The dominant herbaceous species are *Desmostachya bipinnata* and *Clerodendron viscosum*.
2. ***Submontane broadleaf summer deciduous forests*** are found on the nutrient poor and dry sites at the lower altitudinal range around 1000 m. The dominant tree species are *Albizia procera, Terminalia tomentosa, Aldina cordifolia, Toona ciliata, Bauhinia retusa*, and *B. variegata*. The dominant herbaceous species are *Chrysopogon fulvus, Apluda mutica, Arundinella setosa, and Eulaliopsis binata*, etc.
3. ***Low montane needle leaf forests*** are found between the altitudinal ranges of 1000-1800 m. *Pinus roxburghii* (chirpine) is the dominant tree species. *Arundinella pumila, Chrysopogon serrulatus, Botherochloa intermedia, Anthraxon lancefolius, Brachiaria* spp., etc. are the dominant herbaceous species.
4. ***Low to mid-montane hemi-schlerophyllus broadleaf forests*** are distributed between 1500-3000 m altitudes. Tree species of oaks, *viz., Quercus leucotrichophora* (Banj oak) extensively covers areas between 1500 and 2100 m; *Q. lanuginosa* (Rianj oak) occurs in some pockets between 1800-2200 m altitudes; *Q. floribunda* (Tilonj oak) occurs in limited areas between 2000 and 2300 m; and *Q. semecarpifolia* (Kharsu oak) species is found between 2400

and 3000 m. All the oak species are sources of fodder. Canopy of this forest type is close when undisturbed and herb layer is poorly developed with few grasses. In disturbed sites herbaceous plants like *Chrysopogon serrulatus, Arundinella nepalensis, Carex cruciata, C. nubigena, Eulalia mollis, Themeda anathera, Poa annua*, etc. are dominant.

5. ***Mid-montane needle leaf evergreen forests*** occur between 1700 and 3000 m altitudes. The dominant tree species are *Cedrus deodara* (Deodar), *Abies pindrow* (Silver fir), *Picea smithiana* (Spruce), and *Pinus wallichiana*. Herbaceous layer is poorly developed and the dominant species are *Poa annua, Polygonum amplexicaulis, Stellaria hemalayansis, Potentilla* spp., etc.
6. ***Mid-montane winter deciduous forests*** are found between 2300 and 3000 m altitudes. *Acer mono, A. pictum, Aesculus indica, Fraxinus micrantha, Betula alnoides*, and *Caprinus viminea* are the dominant tree species of these forests. *Anemone nepalensis, Esholtzia accuminata, Poa annua, Chrysopogon* spp., etc. are the dominant herbs in these habitats.
7. ***High montane mixed stunted forests*** are found above 3000 m. The dominant canopy trees are *Betula utilis, Abies spectabilis* and the under canopy scrub is *Rhododendron campanulatum*. The herbaceous species are *Potentilla nepalensis, Polygonum amphibium, Trachydium roylei, Geranium wallichianum, Kobresia duthei, Carex alpina*, etc.
8. ***Very high montane scrub*** is dominated by the herbaceous species, like *Danthonia cachemyriana, Potentilla nepalensis, Poligonum amphibium, Trachydium roylei, Geranium wallichianum, Kobresia duthei, Carex alpina*, etc. The dominant woody components are *Rhododendron campanulatum, R. anthopogon*, Salix (less than 1 to 3 m high), etc.

Vegetation and peculiar environmental factors sustain a variety of wildlife since distant past. However, modern developmental activities and denudation of vegetation threatened the majority of wild fauna of many tracts of Uttarakhand mountains. Some of the common wildlife elements are presented in Table 4.1.

Table 4.1: Common Wild Fauna in Uttarakhand

Zoological Names	*Vernaculars*	*English Names*	*Distribution*
	MAMMALS		
Axix axis Erxleben	*Cheetal*	Spotted Deer	Submontane
Axis porcinus Zimmer.	*Pada*	Hog Deer	Submontane
Boselaphus tragocamelus Pallas	*Neel Gai*	Blue-Bull	Submontane
Cannis aureus L.	*Siyar*	Asiatic Jackal	Submontane
Capricornis sumatraensis Bechstein	*Serow*	Serow	Submontane
Cervus duvauceli Cuvier	*Barasingha, Jadau*	Swamp Deer	Submontane
Cervus unicolor Kerr	*Sambar*	Sambar	Submontane-montane
Cuon alpinus Pallas	*Jangli-kutta, Dhole*	Wild Dog	Submontane
Elephas maximus L.	*Hathi*	Indian Elephant	Outer Himalaya
Felis bengalensis Kerr	*Chotta-billi*	Leopard Cat	Submontane
Felis chaus Guldst.	*Jangli-billi*	Jungle-Cat	Submontane
Funambulus palmarum L.	*Gilheri*	Stripped Squirrels	Submontane
Hystrix indica Kerr	*Saula, Sahi*	Porcupine	Submontane-montane
Herpestes auropunctatus Hodgson	*Newla*	Indian Mongoose	Submontane
Herpestes edwardsi Geoffroy	*Newla*	Common Mongroose	Submontane
Hyaena hyaena L.	*Lakadbhagga*	Stripped Hyena	Submontane
Lepus nigricollis F. Cuvier	*Khargosh*	Indian Hare	Submontane
Macaca assamensis True	*Bandar*	Assamese Monkey	Submontane-montane
Macaca mulatto True (Zimmer.) *villosa*	*Bandar*	Monkey	Submontane-montane
Martes martes Bodd.	*Chuthrol*	Pine Marker	Submontane-montane
Melursus ursinus Shaw	*Bhalu*	Sloth Bear	Submontane
Muntiacus muntjak Zimmer.	*Kakar*	Barking Deer	Submontane-montane

Contd...

Table 4.1–Contd...

Zoological Names	*Vernaculars*	*English Names*	*Distribution*
Mus musculus (L.) *homourus* Hodgson	*Musoo*	House Mouse	Submontane
Naemorhedus goral Hardwicke	*Goral*	Goral	Montane
Panthera pardus (L.) *fusca* Meyer	*Bagh*	Panther Leopard	Submontane-montane
Panthera trigis-tigris L.	*Tendu, Bagh*	Tiger	Submontane
Panthera uncia Schreber	*Safedbagh*	Snow Leopard	Montane
Presbytis entellus Dufresne	*Langur*	Common Langur	Submontane-montane
Pteropus giganteus Brunnich		Flying Fox	Submontane
Rhinolopus contrus Temminck	*Chamgadar*	Horse-shoe Bat	Submontane
Selenarctos thibetanus Cuvier	*Reechh*	Himalayan Bear	Montane and above
Sus scrofa (L.) *cristantus* Wagner	*Suar*	Wild Boar	Submontane
Viverricula indica Desmarest	*Malpusa*	Indian Civet	Montane
Vulpes bengalensis Shaw	*Lomdi, Lomri*	Indian Fox	Submontane-montane
	BIRDS		
Acridotheres fuscus Wangler	*Maina*	Jungle Myne	Submontane
Anser poecilorhyncha Penn.	*Gugral*	Spot Bill	Submontane
Aquila heliacea Temm.	*Garur*	Imperial Eagle	Submontane
Aquila rapex Temm.	*Okaw*	Twany	Submontane-montane
Bubulcus ibis Bodd.	*Bagula*	Cattle Egret	Submontane
Columba intermedia Strickl.	*Kabutar*	Blue Rock-Pigeon	Submontane
Columba leuconota Vigors	*Safed Malio*	Snow Pigeon	Montane

Contd...

Table 4.1–Contd...

Zoological Names	*Vernaculars*	*English Names*	*Distribution*
Cissa crythorphyncha Bodd.	*Lampuchiya*	Blue Magpie	Submontane-montane
Coracias bunghalensis L.	*Neelkanth*	Indian Roller	Submontane
Corvus macrorhynchas Adams.	*Kawa*	Jungle Crow	Submontane-montane
Corvus splendens Viell.	*Kawa*	House Crow	Submontane-montane
Coturnix coturnix L.	*Bater*	Common Quail	Submontane-montane
Cuculus saturaths Blyth.	*Kaphal-pako*	Himalayan Cuckoo	Montane
Dendrocygna javanica Horsf.	*Silli, Kath-bau*	Whishling Teal	Submontane
Eudynamys scolopacea L.	*Koyal*	Koel Cuckoo	Submontane
Francholinus francholinus Bons.	*Titar*	Black Patridge	Submontane-montane
Gallus gallus Robin ex Skeels	*Mungi*	Jungle Fowl	Submontane
Glaucidium cuculoides Vigors	*Ullu*	Barred Owlet	Submontane
Glaucidium radiatum L.	*Ullu*	Jungle Owlet	Submontane-montane
Grus antigone L.	*Saras*	Crane	Submontane
Gyps himalayensis Hume.	*Gidh*	Himalayan Vulture	Submontane-montane
Gyps indicus Hume.	*Gidh*	Indian Vulture	Submontane-montane
Haliaestus indus Bodd.	*Cheel*	Brahimini Kite	Submontane-montane
Hypsipetes madagascariensis Miller	*Bulbul*	Black Bulbul	Submontane-montane
Ictinaetus malayensis Tern.	*Cheel*	Black Eagle	Submontane-montane
Lophophorus impejanus Lath.	*Monal*	Monal Pheasant	Montane
Lophura leucomelana Lath.	*Kaleej*	Kallej Pheasant	Montane

Contd...

Table 4.1–Contd...

Zoological Names	*Vernaculars*	*English Names*	*Distribution*
Milvus migrans Skyes	*Cheel*	Black Kite	Submontane-montane
Passer domesticus L.	*Gauraiya*	House Sparrow	Submontane-montane
Passer montanus L.	*Choti Gauraiya*	Tree Sparrow	Submontane-montane
Pavo cristatus L.	*Mor*	Pea Fowl	Submontane
Picus canus Baker	*Kathphoda*	Black-napped Wood Pecker	Submontane-montane
Picus flavinucha Koetz.	*Kathphoda*	Yellow-napped Wood Pecker	Submontane-montane
Picus squamatus Vigors	*Kathphoda*	Scaly Wood Pecker	Submontane-montane
Ploceus philippinus Vigors	*Baya*	Baya Weaver	Submontane
Psittacula alexandri Milur.	*Tota*	Rose Breasted Parakeet	Submontane
Psittacula eupatria Hodgson	*Tota*	Indian Parakeet	Submontane
Spizaetus limnaetus Gmelin	*Shoh-baj*	Hawk	Submontane-montane
Streptopelia decaocto Lath.	*Ghugati*	Ringed Dove	Submontane-montane
Streptopelia orientalis Lath.	*Ghugati*	Rufous Turtle	Submontane-montane
Strix javanica Gmelin	*Ullu*	Javanica Owl	Submontane-montane
Tadorna ferruginea Pallas	*Chakwa*	Brahimini-Duck	Submontane
Tockus birostris Scop.	*Dhanash*	Grey Hornbill	Submontane-montane
Torgus calvus Scop.	*Raj-Gidh*	King-Vulture	Submontane-montane
Treoron apicauda Blyth.	*Hariyal*	Pin-tailed Pigeon	Submontane-montane
Treoron phoenicoptera Lath.	*Malio*	Green Pigeon	Submontane-montane
Turdoides stratus Dumortier	*Saat Bhai, Saat Bahin*	Jangle Babbler	Submontane-montane
Zoothera monticola Vigors		Large Brounthrush	Submontane

Contd...

Table 4.1–Contd...

Zoological Names	Vernaculars	English Names	Distribution
		REPTILES	
Ahaetulla nasutus L.	*Haru-Sanp*	Vine Snake	Submontane
Ancistrodon himalanus Gunt.	*Sanp*	Himalayan Snake	Submontane-montane
Bungnarus caeruleus Dand.	*Krait*	Common Krait	Submontane-montane
Bungnarus fasciatus L.	*Krait*	Banded Krait	Submontane-montane
Callotes versicolor L.	*Chipara*	Garden Lizard	Submontane-montane
Crocodilus poresus Lers.	*Magar*	Crocodile	Outer Himalaya
Elaphe helena L.	*Bhurya Sanp*	Trinket Snake	Submontane
Gavialis gangeticus Gray	*Gadiyal*	Crocodile	Submontane
Hemidactylus brooki Gray	*Chipkali*	House Gecko	Submontane-montane
Naja naja Schleg.	*Nag*	Indian Cobra	Submontane-montane
Ophiophagus . Hannah Schleg	*Kala Nag*	King Cobra	Submontane-montane
Python molurus Gray	*Ajgar*	Python	Submontane
Ptyas mucosus Cope.	*Dhaman*	Rat Snake	Submontane
Varanus monitor L.	*Golu*	Indian Monitor Lizard	Submontane-montane
Sitana pouticeriana L.	*Chiparu*		Submontane
Spalerosophis diadema L.	*Sanp*	Royal Snake	Submontane
Xenochropis piscator L.	*Gadherya-Sanp*	Water Snake	Submontane-montane
		FISHES	
Acrossocheilus hexagolepis Buch.-Ham.	*Kateli*		Submontane
Bagarius bagarius McClelland	*Guch*	Giger Fish	Submontane
Barilius bendelisis Hamilton	*Dhaur*	Hill Trout	Submontane-montane
Barilius bola Lath.	*Bola*	Hill Trout	Submontane-montane

Contd...

Table 4.1–Contd...

Zoological Names	*Vernaculars*	*English Names*	*Distribution*
Crossocheilus latius Hamilton		Sunhera	Submontane
Garra gotyla gotyla Gray	*Gondal*		Submontane
Garra lamta Hamilton	*Gondal*		Montane
Glyptothorax cavia Hamilton	*Siplya*		Submontane-montane
Labeo dero Hamilton	*Kharont*	Rohu	Submontane
Labeo rohita Gunt.	*Rohu*	Rohu	Submontane
Mastacembelus armatus Lac.	*Gair*	–	Submontane-montane
Noemacheilus zonatus McClelland	Gadiyal		Submontane-montane
Pseudechenseis sulcatus McClelland	*Mungria*		Submontane-montane
Salmogairdnerii gairdnersi Lac.	*Salmon*	Gairie	Submontane-montane
Schizothorax richardsonii Gray	*Asalain*	Snow Trout	Submontane-montane
Schizothorax sinuatus Hackel	*Asalain*	Snow Troui	Submontane-montane
Schizothoraichthys progastus McClelland	*Asalain*	Snow Trout	Submontane-montane
Tor tor Hamilton	*Mahseer*	Mahseer	Submontane-montane
Tor putitora Hamilton	*Mahseer*	Mahseer	Submontane-montane

Source: Gaur (1999)

Alpine meadows in the Greater Himalaya are ideal habitat for blue sheep (*Pseudois nayaur*). They are well adapted to semi-arid, moderate to very steep rolling and broken alpine terrain near rugged cliffs (Schaller 1977, Wilson 1981). As herbivores, they feed on a wide variety of grasses and other herbaceous plants, as well as forbs and shrubs. Grasses, however, are the preferred plants on which blue sheep like to relish the most. During summer, they would graze freely over wide range, but seldom more than 200-400 m away from escape cover (Wegge 1979). In winter months, however, they congregate on snow-free slopes and forage on whatever the vegetation is available, primarily residual shrubs and forbs (Richard 1994). Some of the forage species of blue sheep are given in Table 4.2.

Table 4.2: Forage Species for Blue Sheep in the Greater Himalaya

Shrubs	*Berberis sp.*
	Caragana brevifolia
	Catoneaster microphyllus
	Ephedra gerardiana
	Juniperus sp.
	Lonicera spinosa
Forbs	*Anaphalis contorta*
	Polygonum spp.
	Rumex nepalensis
	Thermopsis barbata
	Fern
Grasses and Sedges	*Agrostis pilosula*
	Arundinella nepalensis
	Deschampsia caespitosa
	Danthonia schneideri
	Festuca sp.
	Poa pagophila
	Trisetum sp.

Source: Richard (1994).

South-facing slopes of the mountains are more preferred by black sheep, especially when the temperature is low (Wegge 1979, Prieme and Oksnebjerg 1992). Snow-free ground, especially in summer months, increases thermal stress and escalates vulnerability to predators, lambs and yearlings being the most vulnerable to such mortality (Richard 1994).

Snow leopards (*Panthera uncia*) are found on higher elevations in the Greater Himalaya. The habitat is common with the blue sheep. Blue sheep is its primary prey. They seldom venture to the lower altitudes where livestock graze during grazing season. They, therefore, hardly pose any predation problem to the domestic animals. Their population is directly proportional to that of their preys, the blue sheep, as is the case in the Sagarmatha area of east Nepal (Jackson and Ahlborn 1988). Snow leopard numbers are greatest in terrain broken by cliffs, gullies, and rocky outcrops, features that are well distributed in the upper alpine area (Richard 1994).

Wolves (*Canis lupis*) are also found in this habitat. They are the wild animals that pose greatest threat to livestock. They are more successful hunters of livestock than snow leopards as they hunt in packs of two or more (Richard 1994).

Wild dogs (*Cuon alpinus*) are found in wide range of habitats and also pose severe threat to domestic animals. They generally confine to lower altitudes and prey upon domestic animals rather than blue sheep.

Endemism and Extinction

There are two categories of endemics, namely, i) those that are relict of the last remnants of an old taxon whose distribution has shrunk, and ii) those that are of very recent origin which have not yet extended their range. It appears that, in the Himalaya, many endemics are of the second category. Endemism in Himalayan biota is related to a number of physical and biological factors. There is considerable diversity in geology, geography, soil, and climate, giving rise to many macro–and micro-habitats. The diversity in flora and fauna is particularly apparent in the vegetables and faunal types. Three types of factors underlie diversity (after Khoshoo 1995):

1. Transition from cold desert conditions in the western sector of Ladakh to the cool and moist conditions prevailing in Arunachal Pradesh in the eastern sector;
2. Transition from near tropical/ subtropical conditions on the south side of the Himalaya adjoining the Indus-Gangetic Plains, to the cold dry Tibetan Plateau in the north.
3. Transition along the vertical gradient from tropical/ subtropical conditions at lower altitudes to alpine vegetation and finally to arctic conditions at high altitudes.

There are, thus, three main factors that operate simultaneously: (*i*) longitude (west to east), (*ii*) latitude (south to north), and (*iii*) altitude (lower to higher). There is parallelism in vegetation types on account of latitude and altitude. What grows at higher altitudes at lower latitudes often grows at lower altitudes at higher latitudes. This brings about a great deal of complexity in distribution patterns due to the latitude, longitude, and altitude of a particular place in the Himalaya, and it also creates many microclimatic situations. The population of individual species that occupies wide ranges often gets broken into smaller isolated units. It is for this reason that there is a higher rate of endemism in the Himalaya (Khoshoo 1995).

Himalayan wildlife is unique and provides one of the most fascinating panoramas of biodiversity. Many of the wildlife species are endemic to the Himalaya. There are at least 17 Palaeoarctic endemic bird species in the Himalaya, which do not have any adjacent relatives (Ali 1981). Two genera are monotypic, *e.g.*, the *Catreus* (chir pheasant) and *Ophrysia* (mountain quail). The *Callacanthis burtony* (carducline finch) is the third endemic. The other prominent Palaeoarctic endemic species in the Himalaya are the wood snipe (*Capella nemoricola*), beautiful nuthatch (*Sitta formosa*), Himalayan pied woodpecker (*Dendrocopus major*), black-throated jay (*Garrulus glandarius*), leaf warbler (*Phylloscopus inornatus*), pied ground thrush (*Zoothera wardii*), crested black tit (*Parus melanolophus*), and two species of bull twits (Ripley 1961 cited by Khoshoo 1995).

Wild mammals are adapted to high altitudes and some are confined to very small areas only. Among these are herbivores like wild goats and wild sheep like the markhor (*Capra falconeri*), Himalayan thar (*Hemitragus jemlahicus*), bharal (*Ovis matura*), urial (*Ovis orientalis*), nayan (*Ovis ammon hodgsoni*), and ibex (*Capra ibex*). These together with the serow (*Capricornis sumatraensis*), goral (*Nemorhaedus goral*), takin (*Budorcus taxicolor*), hangul (*Cervus elaphus hanglu*), shou (*Cervus elaphus wallichi*), musk deer (*Moschus moschiferus*), red panda (*Ailurus fulgens*), wild yak (*Bos mutus*), and others are endangered to varying degrees. The same is true of trans-Himalayan species like the Tibetan wild ass (*Asinus kiang*) and Tibetan Gazelle (*Procapra picticaudata*) and antelope (*Pantholops hodgsoni*) (Khoshoo 1995).

The carnivores that are endemic to the Himalaya and also rare are brown bear (*Ursus arctos isabellinus*), red panda, binturong (*Arcticus binturong*), spotted linsang (*Prionodon pardicolor*), species of fox (*Vulpes*), marten (*Martes foina intermedia, M. flavigula*) and weasel (*Mustels sibirica* and other species), civet (*Viverra zibetha*), wolf (*Canis *lupus*), red fox (*Vulpes vulpes*), several species of cats, *e.g.*, lynx (*Felis lynx isabellina*), clouded leopard (*Neofelis nebulosa*), Pallas cat (*Felis manul*), and snow leopard (Khoshoo 1995).

A number of wildlife species, due to several factors leading to considerable deterioration of their habitats in the Himalaya, are endangered / vulnerable. There are some 69 mammals, 37 bird, three reptiles, and two amphibians that are endangered or vulnerable in the Himalaya (Table 4.3).

Table 4.3: Endangered/Vulnerable Animals in the Himalaya

MAMMALS

Slow Loris (*Nycticebus coucang*), Pig-tailed Macaque (*Macaca nemestrina leonina*), Stump-tailed Macaque (*Macaca arctoides*), Capped Langur (*Presbytis pileatus*), Golden Langur (*Presbytis geei*), Phayre's Leaf Monkey (*Presbytis phayrei*), Hoolock Gibbon (*Hylobates hoolock*), Chinese Pangolin (*Manis pentadactyla aurita*), Indian Pangolin (*Manis crassicaudata*), Wolf (*Canis lupus*), Jackal (*Canis aureus*), Red Fox (*Vulpes vulpes*), Indian Fox (*Vulpes bengalensis*), Tibetan Fox (*Vulpes ferrilatus*), Indian Wild Dog (*Cuon alpinus*), Brown Bear (*Ursus arctos isabellinus*), Himalayan Black Bear (*Selenoroctos isabellinus*), Malayan Sun Bear (*Helarctos malayanus*), Sloth Bear (*Melursus ursinus*), Red Panda (*Ailurus fulgens*), Ermine or Stoat (*Mustela erminea*), Weasel (*Mustela sibirica, M. kathia, M. altaica*), Marten (*Martes foina intermedia, M. flavigula*), Marmot (*Marmota bobak himalayana, M. cauduta*), Ratel or Honey Badger (*Mellivora capensis*), Hog Badger (*Arctonyx collaris*), Spotted Linsang (*Prionodon pardicolor*), Binturong (*Arctictis binturong*), crab-eating Mongoose (*Herpestes urva*), Jungle Cat (*Felis chaus*), Pallas' Cat (*Felis manul*), Golden Cat (*Felis temmincki*), Leopard Cat (*Felis bengalensis*), Rusty Spotted Cat (*Felis rubiginosa*), Fishing Cat (*Felis viverrina*), Marbled Cat (*Felis marmorata charltoni*), Lynx (*Felix lynx isabellina*), Clouded Leopard (*Neofelis nebulosa*), Leopard or Panther (*Panthera pardus*), tiger (*Panthera tigris*), Snow Leopard (*Panthera uncia*), Oriental Small-clawed Otter (*Aonyx cinerea*), Eurasian Otter (*Lutra lutra*), Smooth Indian Otter (*Lutra perspi illata*), Great One-horned Rhinoceros (*Rhinoceros unicornis*), Kiang (*Asinus kiang*), Pygmy Hog (*Sus salvanius*), Musk Deer (*Moschus moschiferus*), Swamp Deer (*Cervus duvauceli*), Manipur Thamin (*Cervus eldi eldi*), Hangul (*Cervus elaphus hanglu*), Shou (*Cervus elephus wallichi*), Gaur (*Bos gaurus*), yak (*Bos mutus*), Wild Buffalo (*Bubalus bubalis*), Tibetan Antelope or Chiru (*Pantholops hodgsoni*), Tibetan Gazelle (*Procapra picticaudata*), Takin (*Budorcas taxicolor*), Goral (*Nemorhaedus goral*), Himalayan Tahr (*Hemitragus jemlahicus*), Ibex (*Capra ibex*), Markhor (*Capra falconeri*), Nayan (*Ovis ammon hodgsoni*), Urial (*Ovis orientalis*), Bharal (*Ovis mahura*), Hispid Hare (*Caprolagus hispidus*), Crestless Himalayan Porcupine (*Hystrix hodgsoni*), Serow (*Capricornis sumatraensis*), Gangetic Dolphin (*Platanista gangetica*).

Contd...

Table 4.3–Contd...

BIRDS

Eastern White Stork (*Ciconia ciconia boyciana*), Large Whistling Teal (*Dendrocygna bicolor*), Pink-headed Duck (*Rhodonessa caryophyllacea*), White-winged Wood Duck (*Cairina scutulata*), Himalayan Golden Eagle (*Aquila chrysaetos daphanea*), Himalayan Bearded Vulture (*Gypaetus barbatus aureus*), Osprey (*Pandion haliaetus haliaetus*), Shahin Falcon (*Falco peregrinus peregrinator*), Pheasant Grouse (*Tetraophasis szechenyii*), Assam Bamboo Partridge (*Bambusicola fytchii hopkinsoni*), Mountain Quail (*Ophrysia superciliosa*), Blood Pheasant (*Ithaginis cruentus*), Western Tragopan (*Tragopan melanocephalus*), Satyr tragopan (*Tragopan satyra*), Blyth's Tragopan (*Tragopan blythii*), Temminck's Tragopan (*Tragopan temminckii*), Himalayan Monal Pheasant (*Lophophorus impejanus*), Sclater's Monal Pheasant (*Lophophorus sclateri*), Elwes's Eared Pheasant (*Crossoptilon harmani*), Koklass Pheasant (*Pucrasia macrolopha*), Cheer Pheasant (*Catreus wallichii*), Hume's Bartailed Pheasant (*Syrmaticus humiae humiae*), Peacock Pheasant (*Polyplectrone bicalcaratum*), Indian Peafowl (*Pavo cristatus*), Black-necked Crane (*Grus nigricollis*), Hooded Cranes (*Grus monacha*), Siberian Crane (*Grus leucogeranus*), Masked Finfoot (*Helipais personata*), Bengal Florican (*Eupodotis bengalensis bengalensis*), Jerdon's Courser (*Cursorius bitorquatus*): Extinct, Great Horned Owl (*Budo bubo*), Hodgson's Frogmouth (*Batrachostomus hodgsoni hodgsoni*), Assam Brownbacked Hornbill (*Ptilolaemus tickelli austeni*), Rufous-necked Hornbill (*Aceros nepalensis*), Assam Wreathed Hornbill (*Rhyticeros undulatus ticehursti*), Indian Pied Hornbill (*Anthracoceros malabaricus malabaricus*), Great Pied Hornbill (*Buceros bicornis homrai*).

REPTILES

Gharial (*Gavialis gangeticus*), Mugger (*Crocodylus palustris*), Reticulated Python (*Python reticulatus*).

AMPHIBIANS

Himalayan Newt (*Tylototriton verrucosus*), Garo Hills' Tree Toad (*Nectophryne kempi*).

Source: Khoshoo (1995).

There is no authentic study that could bring out list of animals that are already extinct. However, WCMC (1992) has the following record of extinction of one mammal and three birds in the Himalaya (Khoshoo 1995):

1. The lesser one-horned rhinoceros (*Rhinoceros sondaicus*) was once known to extend from Bengal, Assam, and south of the Brahmaputra in India to Sumatra, Java, and Borneo. At present, it is extinct in India and occurs only in Java.

2. The pink-headed duck (*Rhodonessa caryophyllacea*) was last recorded in the Indian and Nepal Himalaya in 1935 and the cause of extinction has been over-hunting.
3. The Himalayan mountain quail (*Ophrysia superciliosa*) was last recorded in the Himalaya in 1968 and the cause of its extinction has been over-hunting.
4. The forest owlet (*Athene blewitti*) was last recorded in the Himalaya in 1914.

Wildlife Habitat Conservation

Himalayan mountains are in the throes of ecological crises most of which arise due to erroneous, short-term, and insensitive policies and development programmes. Mega hydro projects, indiscriminate construction works, swelling urban areas, tourism, etc. are the major reasons for reckless deforestation and deteriorating wildlife habitats. Poaching, hunting, rampant trade, increasing pollution, etc. are the other reasons to bring immeasurable losses to wildlife.

National Parks, Wildlife Sanctuaries, and Biosphere Reserves are of great significance in conserving wildlife but still inadequate measures to ensure conservation of wildlife in its full diversity. Many wild species prosper in very isolated smaller areas associating themselves with specific plant species and ecological niches. National Parks, Wildlife Sanctuaries, and Biosphere reserves may not represent all the crucial ecological niches or habitats. Therefore, conservation of all the vegetation types that make wildlife habitats and ecological niches for specific animal species is absolutely necessary. In Uttarakhand such protected areas include one Biosphere Reserve and World Heritage Site (Nanda Devi), six National Parks and six Wildlife Sanctuaries located in different biogeographical zones, which are serving the purpose of conservation of important biotypes.

Again, *ex-situ* conservation of wild animals will only accentuate the process of their extinction. Our imperative has to be in-situ conservation of wildlife, for it is the conservation in natural habitats, as a system, rather than as separate entities. Complementarity, mutualism, symbiosis, synergy, and co-evolution of all the associated species in a specific habitat will also be ensured through in-situ conservation. Wildlife can be saved, conserved, and allowed to further flourish by withdrawing and curbing the factors and reversing the processes that lead to deterioration and loss of wildlife habitats. Conservation of wildlife habitats is the most essential process for protection and conservation of wildlife in its magnificent biodiversity. Our main thrust to conserve wildlife–which is must for the essential ecological integrity of our production and natural systems–should, therefore, be on conserving diverse wildlife habitats with their natural biodiversity.

Chapter 5

Utilisation Pattern of Some Important Fodder Resources

Conventional animal husbandry practices in the less developed and developing countries largely built around the so-called White Revolution technology give more emphasis on cultivating fodder crops on arable land in order to increase animal production. Many regions of the world, however, do not spare considerable area for cultivation of fodder crops as part of animal husbandry. In India, for example, not more than four per cent of the arable land is devoted to raising fodder crops. Smallholders comprising the majority of livestock farmers in the mountain areas prefer raising crops for their own consumption and are not in a position to make animals competent for the limited land they own. Only dry fodder (residues of food grain crops) and some fodder trees are raised on the agroforestry systems, which are cultivated areas. Even the large holders find cereal and other cash crops far more remunerative than the fodder crops (Singh *et al.*, 1995).

It should also be realized that in many parts of India animals feed more on rangeland based fodders–trees, shrubs, grasses and other herbs–than on cultivated fodders. Mountain areas of Jammu and Kashmir, Himachal Pradesh, Uttarakhand, Darjeeling district of West Bengal, Sikkim and Arunachal Pradesh in the Indian Himalaya and hilly areas of North-East are such a region where uncultivated rangelands are the main source of livestock feed.

Commons and Private Property Resources of Fodder

For the purpose of convenience, the land use systems can be divided into two categories–common property resources (CPRs) and private property resources (PPRs)–on which livestock owners depend for raising their livestock. CPRs include forest areas/ rangelands, particularly the community forests, marginal lands, and other spaces, such as roadsides, riversides, etc. PPRs include croplands (including agroforestry systems), private grassplots, homestead based fodders, kitchen wastes, etc. Crop residues and grass hay comprise the dry fodders fed to livestock. Crop residues of cereals and pulses are produced on private cultivated land while the grass hay comes from the enclosed areas in the rangelands. Some of the major fodder resources associated with the two types of land use categories are listed in Table 5.1.

Contribution of Different Feed Sources

It was estimated that, on an average, some 82 per cent of dry fodder in the form of crop residues is grown on cultivated land and only 18 per cent comes as grass hay from rangelands. The cultivated land and the rangelands constitute private property resources (PPR) and common property resource (CPR), respectively. Maximum amount of dry fodder (30 per cent) is contributed by wheat straw, followed by barnyard millet (18 per cent), grass hay (18 per cent), finger millet (16 per cent), rice (14 per cent), barley and pulses straw (about 2 per cent each) (Figure 5.1).

Findings corroborate well with those of Singh and Naik (1997), Singh (2000) and Bohra and Singh (2005). Millet straws, especially the barnyard millet straw, are the most preferred ones by the livestock farmers.

Table 5.1: Main Feed Resources in the Himalayan Mountains

Feed	*Common Property Resources*	*Private Property Resources*
Main trees/ shrubs	Oak (*Quercus leucotrichophora*)	Bhimal (*Grewia optiva*)
	Genthi (*Boehmeria rugulosa*)	Kharik (*Celtis australis*)
	Syanru (*Debregeasia salicifolia*)	Bedu (*Ficus* spp.)
	Rampauren (*Vitis flexuosa*)	Kuaral (*Bauhinia* spp.)
	Kaul (*Machilus duthiei*)	Bakain (*Melia azedarach*)
	Saujan (*Ougeinia dalbergioides*)	Tun (*Toona ciliata*)
Crop residues/ dry fodders	Grass hay	Wheat (*Triticum aestivum*)
		Rice (*Oryza sativa*)
		Barnyard millet (*Echinochloa* spp.)
		Finger millet (*Eleusine coracana*)
		Pulses
Herbaceous plants	Grasses, palatable herbs	Grasses on field bunds
		Weeds
Concentrate		Wheat flour
		Barnyard millet grains
		Finger millet flour
		Soybean
		Kitchen waste

Farmers in Uttarakhand Himalaya, as virtually in all parts of the Himalaya, overwhelmingly depend on CPRs for livestock raising. This is one of the unique features of mountain areas' land-based economies. Size of the holdings, therefore, is not a constraint to livestock husbandry in mountain areas as could be the case in the plains where CPRs have become redundant or a thing of the past. CPRs are the major source of green fodders–tree / shrub leaves, green grasses and other herbaceous plants. PPRs provide bulk of the dry fodders in the form of crop residues. Tree / shrub leaves, grasses from

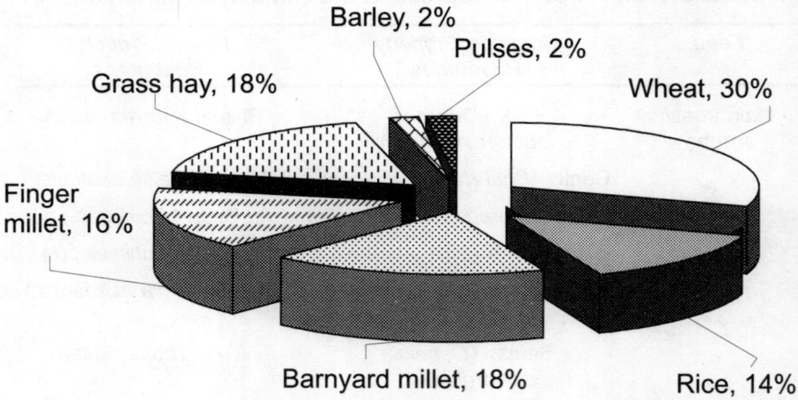

Figure 5.1: Proportion of Different Sources of Dry Fodder Contributing to Livestock Feeding

the field bunds and weeds extracted from the crop fields form the greens available in the PPR-based land use.

Green grass (hand-cut) accounts for 30 per cent of the total greens, while grazed grass as high as 55 per cent. Tree leaves amounted to only about 12 per cent. The other green comprising seasonal crops such as mustard, field weeds, etc. constitute about 3 per cent of the total green fodder in a year, on an average (Figure 5.2). While virtually entire grazed grass is the product of the rangelands, most of the hand-cut grass (including weeds growing in the fields) comes from the bunds of the terraced fields in the cultivated land area. Out of 12 per cent biomass of tree leaves, about 5 per cent comes from the rangelands and the rest from the croplands. Figures obtained in this study are in close proximity with those obtained by Singh and Naik (1987), Singh (2000) and Bohra and Singh (2005).

Most of the concentrate feed utilized by dairy animals is home-produced. Barnyard millet, barley, finger millet flour, white-seeded soybeans, black soybeans, wheat bran, wheat flour, etc. make the home-grown concentrate feed. The home-produced concentrate is often fed to cows and buffaloes when they are in milk and to bullocks when they are at work. The dairy animals when dry and bullocks during idle days are not given concentrate.

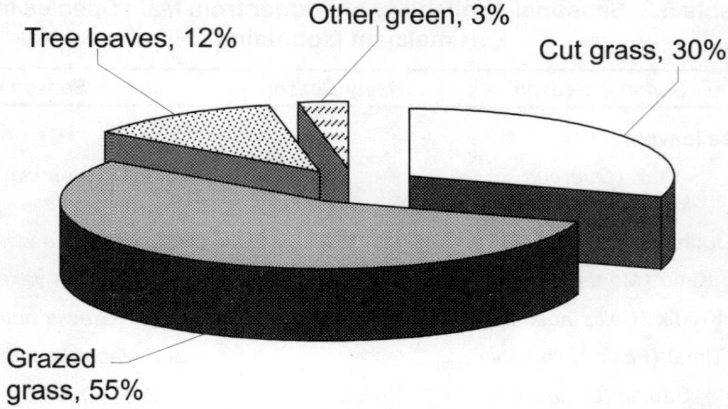

Figure 5.2: Proportion of Different Sources of Green Fodder Utilized for Livestock Feeding

Seasonal Availability

All fodder-yielding species are not available throughout the year. Seasonality of fodder availability is one of the important issues facing livestock farmers in the mountains. Provision of feeds and fodders to livestock in mountain areas would largely depend on the type of an agro-ecological zone.

The feeds and fodders presented in Table 5.2 are only those on which the animals are largely dependent. *Quercus leucotrichophora, Ficus* spp., *Morus serrata, Celtis australis, Ougeinia dalbergioides, Machilus duthiei*, etc. are the main fodder trees that are harnessed during summer season. *Ficus* spp., *Bauhinia* spp., *Grewia optiva, Machilus duthiei, Quercus leucotrichophora, Vitis flexuosa*, etc. are the major fodder trees topped during winter season. Stinging nettle plant is given only during winter months, particularly to buffaloes. Rainy season helps to produce plenty of green grasses and other herbaceous plants which are utilized as fodder. Dry fodders (crop residues and grass hay) are often fed to livestock during winter season. Concentrate, mostly home-produced, is given throughout the year. It is generally given to milking animals only.

Table 5.2: Seasonal Availability of Fodder from Main Species in the Himalayan Mountains

	Summer Season	*Rainy Season*	*Winter Season*
Tree leaves			
	Oak (*Quercus leucotrichophora*)		Chanchri (*Ficus clavata*)
	Dudhla (*Ficus nemoralis*)		Malu (*Bauhinia vahlii*)
	Kimu (*Morus serrata*)		Kwaral (*Bauhinia variegata*)
	Kharik (*Celtis australis*)		Bhimal (*Grewia optiva*)
	Timal (*Ficus roxburghii*)		Kaul (*Machilus duthiei*)
	Saujan (*Ougeinia dalbergioides*)		Oak (*Quercus leucotrichophora*)
	Kaul (*Machilus duthiei*)		Khinyo (*Ficus cunia*)
			Rampauren (*Vitis flexuosa*)
Other green fodder			
	Field weeds	Field weeds	Field weeds
			Mustard (*Brassica* spp.)
			Stinging nettle (*Urtica dioica*)
Grasses/Herbs			
	Green grass	Green grass	
	Gajao grass (*Themeda anathera*)	Gajao grass (*Themeda anathera*)	
Dry fodder			
	Rice straw	Wheat straw	Rice straw
			Finger millet straw
			Grass hay
			Barnyard millet straw
Concentrate			
	Home grown	Home grown	Home grown

An understanding of the seasonal availability of fodders is important for addressing the issue of seasonality of fodders. Nomads and transhumant pastoralists take full advantage of seasonality by moving their large herds to the place where fodder is available aplenty in different seasons. During summer, domestic animals are herded in the alpine meadows which become lush green after snow melt. Lower altitudes are destitute of fodder during summer. Most of the farming families following sedentary system of livestock keeping in many parts of Uttarakhand move bag and baggage to summer camps situated amidst evergreen forests at high altitudes. This is a way to respond to seasonal fodder scarcity at lower altitudes. High altitude areas have plentiful fodder in the evergreen forests, like those of the oaks, at high altitudes in summer. In winter season, fodder scarcity is responded by feeding grass hay which was prepared at the end of rainy season. Weed plants extracted from crop fields in the winter season are supplemented with livestock feeds, which also respond green fodder scarcity in the season.

Chapter 6
Herbaceous Cover in Range Ecosystems

With increasing pressure for fuel and fodder on the forests, a variety of grassland types have been developed particularly on mid-hills (Yadav and Singh 1977, Melkania and Singh 1989). The vegetational analysis of the community was done in such grass-dominated rangelands to consider only the herbaceous plants of crucial fodder value. Vegetative propagation, easy dispersal of seeds by animals, livestock grazing pressure, inherent natural factors, and human management are the main reasons for the dominance of grass species in the rangeland ecosystems.

This Chapter attempts to analyse the frequency, density, abundance, importance value index and productivity of the species available in the grass-dominated rangeland in the Uttarakhand Himalaya. For this study, quadrats of desired size were used. Methods employed by Mishra (1968), Odum (1971) and Singh and Yadav (1974) were consulted.

Frequency

Frequency as introduced by Raunkiaer (1934) indicates the number of sampling units in which a given species occurs, and thus expresses the distribution or dispersion of various species in a community. From there, per cent frequency is calculated as follows:

$$\text{Frequency\%} = \frac{\text{Number of sampling units in which the species occurred} \times 100}{\text{Total number of units studied}}$$

Required number of quadrats of appropriate size are laid in the rangelands to be studied and noted the presence and absence of each species by + or–respectively in each of the quadrats (Mishra 1968).

Density

The term density represents the numerical strength of species in the community. The density and frequency taken together are of prime importance in determining community structure (Oosting 1958).

A number of quadrats of desired size are laid and the individuals of each species in each quadrat counted. The total number of quadrats studied are recorded. From this, density is calculated as follows:

$$\text{Density} = \frac{\text{Total number of individuals of the species}}{\text{Total number of quadrats studied}}$$

Abundance

Abundance is described as the number of individuals per quadrat of occurrence. The abundance is calculated as follows:

$$\text{Abundance} = \frac{\text{Total number of individuals of the species}}{\text{Total number of quadrats of occurrence}}$$

Importance Value Index (IVI)

IVI of a species is the relative contribution of the species to the whole community. IVI is defined as the "sum of relative density, relative frequency and relative abundance" as:

$$IVI = RD + RF + RA$$

An Analysis of the Herbaceous Species

In the grass-dominated Himalayan rangelands, as in case of the parts of Uttarakhand Himalaya, *Themeda anathera* is found occurring with the average density of 72.5 per m^2, while *Cassia mimosoides* was observed with the least density of 0.5 per m^2. *Digitaria stricta, Pennisetum flaccidum* and *Setaria homonyma* also recorded quite high density (Table 6.1).

On the basis of frequency values various species can be distributed into five frequency classes, *viz.*, A (0-20 per cent), B (21-40 per cent), C (41-60 per cent), D (61-80 per cent) and E (81-100 per cent), as per Raunkiaer (1934). According to this criterion, out of a total number of species recorded, 11 belong to A, 12 to B, 6 to C, 1 to D and 0 to E. Of the total number of species, the representation of each of the five frequency classes is A=36.67 per cent , B=40.00 per cent , C=20.00 per cent , D=3.33 per cent and E=0 per cent . Thus, from these observations, *Setaria homonyma* and *Themeda anathera* are the most frequent species in the grass-dominated rangelands in the Himalayan mountains in Uttarakhand.

Themeda anathera was the dominant grass in the rangelands of the Uttarakhand Himalaya with highest importance value index (IVI) of 43.19, followed by *Digitaria stricta, Pennisetum flaccidum, Setaria homonyma, Erigeron karvinskianus, Echinochloa crus-galli, Eleusine indica, Chrysopogon gryllus, Eragrostis atrovirens, Cyperus rotundus, Sporobolus diander, Oplismenus compositus, Arundinella nepalensis, Digitaria cruciata, Cynodon dactylon, Miscanthus nepalensis, Chenopodium ambrosioides, Agrostis pilosula, Cyperus nutans, Achyranthes bidentata, Pollinia quadrinervis, Dactyloctenium aegypticum, Justicia simplex, Brachypodium sylvaticum, Arthraxon prionodes, Leucas lanata, Oxalis corniculata, Rumex hastatus, Dichanthium annulatum* and *Cassia mimosoides*.

Table 6.1: Vegetational Analysis of Herbaceous Layer of the Rangelands

Species	*Density (per m²)*	*Frequency (F) (%)*	*Abundance (A)*	*A/F*	*IVI*
Themeda anathera	72.5	60	30.20	0.50	43.19
Digitaria stricta	32.4	50	16.20	0.32	22.94
Pennisetum flaccidum	20.2	50	10.10	0.20	16.28
Setaria homonyma	20.2	70	7.21	0.10	17.09
Erigeron karvinskianus	15.5	20	19.37	0.96	15.72
Echinochloa crus-galli	15.2	40	9.50	0.23	13.34
Eleusine indica	14.2	50	7.10	0.14	13.00
Chrysopogon gryllus	13.0	30	10.83	0.36	12.16
Eragrostis atrovirens	10.5	30	8.75	0.29	10.43
Cyperus rotundus	10.2	40	6.37	0.15	10.33
Sporobolus diander	8.2	20	10.25	0.51	9.31
Oplismenus compositus	7.2	50	3.60	0.07	9.18
Arundinella nepalensis	7.2	40	4.50	0.11	9.53
Digitaria cruciata	7.0	50	3.50	0.07	9.07
Cynodon dactylon	5.5	20	6.87	0.34	6.92
Miscanthus nepalensis	5.0	40	3.12	0.07	7.04
Chenopodium ambrosioides	4.5	30	3.75	0.12	6.26
Agrostis pilosula	4.5	30	3.75	0.12	6.26
Cyperus nutans	4.5	10	11.25	1.12	7.40
Achyranthes bidentata	4.2	30	3.50	0.11	6.05
Pollinia quadrinervis	4.2	10	10.50	1.05	7.01
Dactyloctenium aegypticum	4.0	30	3.33	0.11	5.79
Justicia simplex	3.6	10	9.00	0.90	6.18
Brachypodium sylvaticum	3.5	30	2.91	0.09	5.34
Arthraxon prionodes	3.0	20	3.75	0.18	4.60
Leucas lanata	2.8	10	7.00	0.70	4.54
Oxalis corniculata	2.6	20	3.25	0.16	3.70
Rumex hastatus	2.0	30	1.66	0.05	4.52
Dichanthium annulatum	1.2	10	3.00	0.30	2.76
Cassia mimosoides	0.5	20	0.62	0.03	2.53

Bhatt (2005) in his study on the vegetational analysis of a biotic community in Kumaun Himalaya observed that the *Themeda anathera* had the highest density and IVI, which corroborated with this study. Many values in his analysis indicated much similarity with that of ours. However, figures relating to community structure and species diversity of a tropical forest (Tripathi *et al.*, 2004) did not corroborate with the figures emerged in this study. This might be due to basic difference in the community structures in a temperate and a tropical region. Community characteristics, in fact, would vary from area to area and from one rangeland to the other. The characteristics, of course, would also vary from time to time at the same location.

Herbage Productivity

Herbage productivity is one of the key indicators of the performance of a rangeland ecosystem. The average herbage productivity values in different rangelands–*viz.*, oak type, mixed type, oak-grass type, oak-conifer type and grass type–was found to be 121.64±12.92g/ m^2, 78.09±16.19g/ m^2, 153.64±23.89g/ m^2, 128.88±20.66g/ m^2 and 106.00±10.52g/ m^2, respectively in summer season. In rainy season, productivity was substantially more in comparison to that in the summer season, which was recorded to be 501.28±66.65g/ m^2, 763.20±12.88 g/ m^2, 454.68±27.05 g/ m^2, 692.12±81.48 g/ m^2 and 766.88±23.07g/ m^2 in oak type, mixed type, oak-grass type, oak-conifer type and grass type rangelands, respectively. Highest productivity was recorded in oak-grass type rangeland in summer season and in grass type in rainy season.

In winter season, the herbage productivity is critically low and ground flora, especially in grass type and conifer type rangelands, almost dries up. In this season, livestock farmers depend only on grass hay and other crop residues for feeding their livestock. Figures for winter season, therefore, have not been included. Herbage productivity is low in summer season and high in rainy season, which is due to the environmental factors which are unfavourable for plant growth during summer season (Table 6.2).

There is a phenomenal increase in herbage productivity of different rangelands upon the onset of the monsoon season. As high as 623 and 777 per cent increase in herbage productivity has been recorded in the grass type and mixed type grasslands in the rainy season over the summer season. The increase in herbage output ranges from 3 times in oak-grass type rangeland to as many as 10 times in case of mixed type rangeland in the monsoon season than in the summer season.

Table 6.2: Herbage Productivity per m^2 in Different Rangelands in Parts of the Uttarakhand Himalaya During Summer and Rainy Season

Type of Rangeland	*Average Productivity, g per m^2 in Summer Season*	*Average Productivity, g per m^2 in Rainy Season*	*Increase in Productivity Over the Season, %*
Oak type	121.64 ± 12.92	501.28 ± 66.65	312.12
Mixed type	78.096 ± 16.19	763.20 ± 12.88	777.21
Oak–grass type	153.64 ± 23.80	454.68 ± 27.05	195.94
Oak–conifer type	128.88 ± 20.66	692.12 ± 81.48	437.03
Grass type	106.00 ± 10.52	766.88 ± 23.07	623.47

According to Sundriyal (1995), the productivity of the herbage is influenced by seasonality of climate, and biotic pressure on them. Grasslands and forest grazing lands meet about 50-60 per cent of total fodder need in the Himalaya; however, their composition and net primary productivity vary from area to area due to the variation in the species dominance, topography, rainfall pattern, biotic pressure and successional status of grasses.

Chapter 7

Chemical Composition of Fodders

Rangelands harbour a number of plant species many of which are of high fodder value. These ecosystems support virtually entire livestock populations in the Himalayan region. Certain plant species–trees, shrubs and herbs–are of critical importance for livestock production. It is, therefore, crucial to evaluate different plants for their nutritive values. The chemical composition of plants is known to determine the nutritive value of the fodder the plants serve. This Chapter presents the chemical analysis and discusses nutritive values of some fodder-yielding species flourishing in the Himalayan rangelands.

All the fodder samples derived from the plants explored in different types of the rangelands in different parts of Uttarakhand mountains were taken to the laboratory for detailed inventory. The plant samples were analysed for their proximate principles, *viz.*, Dry Matter (DM), Crude Protein (CP), Ether Extract (EE), Crude Fiber (CF), Nitrogen Free Extractives (NFE), Total Ash (TA) and Acid Insoluble Ash (AIA) by the standard methods of AOAC (1995).

Neutral Detergent Fibre (NDF) and Acid Detergent Fibre (ADF) were determined using methods employed by Goering and Van Soest (1970).

Available fodders comprising crop residues and samples of home-produced concentrate (*chaata*) and commercial feeds, *viz.*, complete feed blocks (CFB) and urea-molasses-mineral blocks (UMMB), were also subjected to proximate analysis in order to make comparison between the nutritive values of the rangeland plants and the other available feeds and fodders. While the crop residues, green fodder and *chaata* were collected from the farm families, the commercial feed samples were procured from the GB Pant University of Agriculture and Technology, Pantnagar.

Determination of Dry Matter

Representative samples of rangeland plants and other feeds and fodders were taken in previously weighed silica crucibles / tin trays and kept in hot air oven at 100°C for 24 hours. Dry matter (DM) was calculated as follows:

$$\text{DM per cent} = (B / A) \times 100$$

where,

A: Fresh weight of sample, g

B: Dry weight of sample, g

The dried samples were grounded with a laboratory Willey mill using 2.0 mm and 1.0 mm sieve, mixed well and stored in a clean and well labeled polythene bags for pending chemical analyses. The dry matter determination is the important proximate principle as all the nutritive values and digestibility coefficients are on dry matter basis.

Determination of Crude Protein

Crude protein (CP) in range plants and other feed and fodder samples was estimated by using semi automatic Kjeltec method. Accurately 1.0g samples were taken in Kjeltec digestion tube and digested with 1.0g digestion mixture ($CuSO_4 : K_2SO_4$, 1 : 9) and 12 ml concentrated sulphuric acid. The nitrogen content of the sample was transformed into ammonium sulphate by sulphuric acid digestion. A blank was also prepared by taking all the reagents in digestion tube except the sample. The digested sample of feed and fodder was cooled and subjected to distillation in semi automatic Kjeltec distillation assembly. The ammonia released during distillation was collected into 30 ml of 4 per cent boric acid solution containing mixed indicator (0.2 per cent methyl red and 0.1 per cent bromo cresol green in equal amount in 95 per cent ethyl alcohol). The ammonia collected in boric acid solution was titrated against 0.1 N H_2SO_4. Nitrogen percentage was calculated as follows:

$$\text{Nitrogen per cent} = \frac{(V1\text{-}V2) \times A \times 0.0014}{B} \times 100$$

where,

V1: Volume (ml) of 0.1N H_2SO_4 used for titration of sample

V2: Volume (ml) of 0.1N H_2SO_4 used for titration of blank

A: Dilution factor (volume made by aliquot taken for steam distillation)

B: Weight of sample taken for digestion

0.0014: Molecular weight of nitrogen (g) equivalent to neutralize 1.0ml of 0.1 N H_2SO_4

CP%: Nitrogen per cent × 6.25

Determination of Ether Extracts

Fat and fat soluble components of air-dried feed and fodder samples were estimated by Soxtec Avanti manual system. Solvent extraction in the extraction unit was performed in two steps. First, the sample was immersed in the boiling solvent to dissolve soluble materials. In the second step, the sample was raised above the solvent surface to permit efficient washing with solvent from the condenser.

After the extraction, the condenser valves were closed by lifting the samples to the upper position. After few minutes, most of the solvent was collected via the condenser in a collection vessel. At last, the residue of the solvent was evaporated when the air pump was started. Ether extract was collected in previously weighed extraction cups and weighed after cooling in desiccator. The ether extract (EE) was calculated as follows:

$$\text{EE per cent} = (C - A / B) \times 100$$

where,

A: Initial weight of extraction cups (g)

B: Weight of sample (g)

C: Weight of extraction cups with ether extract (g)

Determination of Crude Fibre

The samples of feed and fodder after de-fatting as mentioned above were transferred from thimbles to spout less beaker of one litre capacity and in each beaker, 200 ml of 1.25 per cent H_2SO_4 was added. It was refluxed for 30 minutes on hot plates after the boiling started and subsequently filtered through muslin cloth. The residue was washed 5-6 times with hot water until it became acid free. The residual materials on the muslin cloth were again transferred to the respective beaker and in each beaker 200 ml of 1.25 per cent sodium hydroxide solution (NaOH) was added. It was refluxed for 30 minutes after the boiling started and subsequently filtered through muslin cloth and washed with hot water for 5-6 times until it became free from alkali. Then, total residue was transferred in a clean dry crucible and dried in hot air oven at 100°C for 24 hours and then it was cooled in a desiccator and weighed. The residue was then ignited in muffle furnace at 600°C for 2 hours. After 12 hours, crucibles containing ash were removed from the furnace and transferred into desiccator, cooled and weighed again. Weight loss due to ignition was recorded as the weight of crude fibre (CF).

$$\text{CF per cent} = (B / A) \times 100$$

where,

A: Weight of sample on DM basis (g)

B: Weight of CF (g)

Determination of Total Ash

Oven-dried 5.0 g sample was taken in a previously weighed silica crucible. The crucible along with the sample was kept on heater and burnt till no smoke was given off by the mass of the sample. With the help of metal tong the silica crucible was transferred into muffle furnace and kept at 600°C for two hours and then transferred into desiccator, cooled and weighed. Total ash (TA) was calculated as follows:

$$\text{Total ash (per cent)} = (A\text{–}B / C) \times 100$$

where,

A: Weight of silica crucible with ash (g)

B: Weight of empty silica crucible (g)

C: Weight of sample taken for ashing (g)

Determination of Organic Matter

Organic matter (OM) was estimated by subtracting per cent ash on dry matter basis from 100 and expressed as per cent on dry matter basis.

$$\text{OM (per cent)} = 100 - [(A - B/C) \times 100]$$

where,

A: Weight of silica dish + ash (g)

B: Weight of empty silica dish (g)

C: Weight of sample taken (g)

Determination of Acid Insoluble Ash

Dilute HCl (1:4) 25 ml was added in the crucible containing ash, covered with watch glass and digested for 20-30 minutes on the hot plate. After digestion, the contents were rinsed with water and filtered repeatedly with 5 per cent HCl solution through Whatman's filter paper no. 42 into a 100 ml volumetric flask.

The filter paper along with the residue left on it was transferred in the same crucible, dried and ignited in a muffle furnace at 600°C for 2 hours. After 12 hours, crucible containing acid insoluble ash was removed from the furnace and transferred into the desiccator, cooled and weighed. The acid insoluble ash (AIA) was calculated as follows:

$$\text{AIA per cent} = (A/B) \times 100$$

where,

A: Weight of acid insoluble ash (g)

B: Weight of sample taken on dry matter basis (g)

Determination of Nitrogen-free Extracts

Nitrogen-free extract (NFE) was arrived at by subtracting the sum of total of crude protein, ether extract, crude fibre and total ash per cent from dry matter per cent. The NFE was calculated as follows:

$$NFE\% = 100 - DM\% + (CP\% + EE\% + CF\% + TA\%)$$

Determination of Neutral Detergent Fibre

Dried sample was ground using 1.0mm sieve and then 1.0g sample was weighed into the spout less beaker. One hundred ml of neutral detergent solution, 0.5g sodium sulphate and 2.0ml of decalene was added into it. The contents of the beaker were then refluxed on hot plate for 60 minutes. Contents were filtered by previously weighed gooch crucible on filter manifold and washed first with hot water 3-4 times and then by acetone twice. The material was transferred into crucible and its contents were dried and weighed. Empty crucible weight was subtracted from the total weight of residue plus crucible. The neutral detergent fibre (NDF) was calculated as follows:

$$NDF\% = \frac{\text{(Weight of crucible + Cell wall)} - \text{Empty weight of crucible}}{\text{Weight of sample}} \times 100$$

Determination of Acid Detergent Fibre

One gram dried sample ground to 1 mm size was taken into spoutless beaker, 100 ml of the acid detergent solution and 2.0 ml of decalene were added. In this case, sodium sulphate is omitted. The rest procedures were same as in NDF determination except that additional washing was done with hexane twice. Acid detergent fibre (ADF) was calculated as follows:

$$NDF\% = \frac{\text{(Weight of crucible + Fibre)} - \text{Empty weight of crucible}}{\text{Weight of sample}} \times 100$$

Determination of Hemicelluloses

The difference between the NDF and ADF content of sample was considered as hemicelluloses.

Hemicelluloses (per cent) = NDF (per cent) – ADF (per cent)

Determination of Dry Matter Digestibility

Digestibility was determined by using nylon bag technique as per the procedure of Mehrez and Ørskov (1977). Dry matter digestibility of feed samples was carried out by using rumen fistulated animals. All the samples were incubated in duplicate bags taken at each incubation. Incubation time was 72 hours. For this technique, bag of 9x17 cm size made of 100 per cent nylon cloth of 40-50 μm pore size were prepared with double stitching and rounded corners. A 5.0g oven-dried and ground sample of each feedstuff was taken in a nylon bag and tied with a nylon thread. The bags were suspended in the rumen of the fistulated animals with the help of long nylon cord for 72 hours. The nylon bags were tied to small iron ring to prevent the floating of bags in the rumen.

After the lapse of stipulation time, the bags were removed from the rumen and washed thoroughly with running water. These residue samples were then dried in an oven at 70°C for more than 48 hours (till constant dry weight was observed). Each bag was opened after drying and the pelleted residue was broken by hand and transferred to labeled and weighed paper bag, and kept further in an oven till constant weight was observed. After that, the residue was quickly weighed. The amount of nutrients before and after the incubation was taken for calculation of per cent disappearances. Dry matter digestibility (DMD) was calculated as follows:

$$\text{Digestibility per cent} = (A - B/A) \times 100$$

where,

A: Weight of sample taken for incubation (g)

B: Weight of sample after incubation

Statistical Analysis

The data relating to chemical composition and nylon bag digestibility of 23 samples were subjected to statistical analysis using Complete Random Design (CRD) as described by Snedecor and Cochran (1968).

Chemical Analysis of Fodders

The samples collected from the fields were immediately weighed and brought to the lab for their chemical analysis. The collected samples were divided into four groups, namely, tree leaves (G1), grasses (G2), crop residues (G3) and commercial feed (G4). This categorization was convenient since it was according to the category of feeds.

Dry Matter

As revealed through Table 7.1, there was a wide variation in chemical composition and nylon bag dry matter digestibility (DMD) of different samples. Average dry matter (DM) content was 36.07±2.88, 36.24 ± 12.44, 89.30±0.59 and 87.13±3.62 per cent in tree leaves, grasses, crop residues and commercial feeds respectively (Table 7.2). Dry matter content varied from 10.58 per cent (*Vitis flexuosa*) to 93.70 per cent (locally prepared concentrate, *Chaata*) in different groups. Among tree leaves, DM percentage was high in *Ficus cunia* (49.96 per cent) and low in *Vitis flexuosa* (10.58 per cent). The other trees had moderate dry matter content. The dry matter content was significantly different ($P<0.01$) among different groups (Table 7.1).

If we compare each group with the other, there was non significant difference between G1 and G2, G3 and G4 but significant difference ($P<0.01$) between G1 and G3, G1 and G4, G2 and G3, G2 and G4 (Tables 7.2, 7.3, 7.4, 7.5, 7.6, 7.7 and 7.8).

Determination of the moisture content in a feed, or dry matter, is essential, for all the nutritive values of a forage or food are considered on dry matter basis.

Crude Protein

Crude protein (CP) includes proteins, amino acids, amines, nitrates, nitrogenous glycosides, glycoproteins, B vitamins and nucleic acids. CP content in a feed is the most important indicator of the quality of feed. Generally, higher the CP content, better the quality of feed for ruminants.

The range of CP in different feeds was from 5.05 per cent (hay of *Themeda anathera*) to 25.05 per cent (*Grewia optiva*) in different groups (Table 7.1). Among the tree leaves, the CP content was highest in *Grewia optiva* (25.05 per cent) and lowest in *Quercus leucotrichophora* (10.77 per cent). The CP content amongst grasses varied from 5.05 per cent (the hay of *Themeda anathera* grass) to 7.40 per cent (mixed green grass). In crop residues, CP content was 3.23 per cent (rice

Table 7.1: Chemical Composition (%) and Nylon Bag Dry Matter Digestibility (DMD, %) of Different Feeds on Dry-Matter Basis

Feeds and Fodders	*DM*	*CP*	*EE*	*CF*	*TA*	*AIA*	*NDF*	*ADF*	*NFE*	*OM*	*HC*	*DMD*
Tree Leaves												
Bhimal (*Grewia opiva*)	42.60	25.05	3.00	16.70	17.51	4.12	46.76	27.78	37.74	82.49	18.98	67.00
Kharik (*Celtis australis*)	33.70	20.39	2.15	15.92	12.57	3.37	55.69	34.95	48.97	87.43	20.74	66.00
Rampauren (*Vitis flexuosa*)	10.58	19.16	2.70	20.17	17.11	1.42	53.25	41.67	40.86	82.89	11.58	63.00
Saujan (*Ougeinia dalbergioides*)	30.65	15.80	1.44	18.58	16.26	0.50	56.52	42.53	47.92	83.74	13.99	40.00
Chanchri (*Ficus clavata*)	40.51	13.37	1.15	17.81	13.63	3.25	54.68	38.15	54.04	86.37	16.53	65.00
Timal (*Ficus roxburghii*)	32.50	15.13	2.70	18.74	17.33	5.85	52.13	36.87	46.10	82.67	15.26	61.00
Beru (*Ficus palmata*)	25.65	17.07	1.08	11.10	12.29	1.74	54.68	30.58	58.46	87.71	24.10	48.00
Kaul (*Machilus duthiei*)	38.53	15.49	1.14	23.04	7.08	0.33	59.70	41.06	53.25	92.92	18.64	49.00
Oak (*Quercus leucotrichophora*)	49.45	10.77	3.81	19.54	4.28	0.56	50.84	42.89	61.60	95.72	7.95	49.00
Dudhla (*Ficus nemoralis*)	34.68	16.53	1.09	17.43	9.15	1.66	53.76	40.23	55.80	90.85	13.53	52.00
Khinyo (*Ficus cunia*)	49.96	12.14	1.78	19.12	8.10	0.88	54.67	39.08	58.86	91.90	15.59	55.00
Malu (*Bauhinia vahlii*)	42.70	16.34	1.21	20.53	15.15	5.88	56.70	43.45	46.77	84.85	13.25	62.00
Panyyan (*Prunus cerasoides*)	37.49	19.15	1.40	9.89	15.87	6.51	53.08	29.02	53.69	84.13	24.06	55.00

Contd...

Table 7.1–Contd...

Feeds and Fodders	*DM*	*CP*	*EE*	*CF*	*TA*	*AIA*	*NDF*	*ADF*	*NFE*	*OM*	*HC*	*DMD*
Grasses												
Themeda anathera green	19.89	5.62	2.33	34.15	6.06	2.35	59.64	34.21	51.84	93.94	25.43	55.00
Mixed green grass	21.97	6.50	1.67	28.09	10.66	5.31	58.20	46.53	53.08	89.34	11.67	48.00
Mixed green grass	25.89	7.05	2.09	26.89	10.17	6.06	52.64	35.29	53.80	89.83	17.35	43.00
Mixed green grass	27.78	7.40	2.91	26.41	12.52	5.20	54.29	37.05	50.76	87.48	17.24	51.00
Themeda anathera hay	85.71	5.05	1.69	30.97	7.05	4.70	54.40	46.82	55.24	92.95	17.47	40.00
Crop Residues												
Rice straw	89.90	3.23	1.45	45.60	10.35	6.06	70.12	42.85	39.37	89.65	27.27	52.00
Finger millet straw	88.70	3.60	1.56	48.89	10.27	4.60	67.02	44.07	35.68	89.73	22.95	52.00
Commercial Feeds												
UMMB	81.20	22.14	0.87	4.87	20.44	7.43	20.87	12.44	51.68	79.56	8.43	85.00
CFB (Complete feed block)	86.50	16.64	2.01	21.69	19.10	7.03	57.87	34.48	40.56	80.90	23.39	68.00
Home-produced concentrate (*Chaata*)	93.70	16.19	6.03	5.17	2.89	0.93	14.60	9.10	69.72	97.11	5.5	90.00

straw) and 3.60 per cent (finger millet straw) and in commercial feeds CP content was highest in UMMB (22.14 per cent) and lowest in *Chaata* (16.19 per cent) (Table 7.1). Average CP content was 16.64±1.03 per cent in tree leaves, 6.32±0.43 per cent in grasses, and 3.14±0.185 per cent in crop residues and 18.32±1.91 per cent in commercial feeds (Table 7.2). Among the groups, CP content was significantly different ($P<0.01$). There was significant difference ($P<0.01$) between group G1 and G2, G1 and G3, G2 and G4, G3 and G4, but the difference was not significant between G1 and G4, G2 and G3 (Tables 7.2, 7.3, 7.4, 7.5, 7.6, 7.7 and 7.8).

CP values of some of the fodder species determined in this experiment corroborate well with those evaluated by Singh *et al.* (1988) and Yadava *et al.* (2003). However, CP values of all the range-based fodder species brought to the fore through this experiment are not available in the literature published earlier and hence cannot be compared.

Ether Extract

Ether extract (EE) fraction in chemical composition of the fodder samples comprises fats, oils, waxes, organic acids, pigments, sterols, vitamins A, D, E and K. These fractions are soluble in organic solvents used in this experiment.

The EE content was highest in *Chaata* (6.03 per cent) and lowest in *Beru* (*Ficus palmata*) (1.08 per cent). In different tree leaves, the EE range was between 1.08 per cent (*Ficus palmata*) to 3.00 per cent (*Grewia optiva*). Among the grasses, the EE range was from 1.67 per cent (*Themeda anathera* green) to 2.91 per cent (mixed green grass). EE content was higher in finger millet straw (1.56 per cent) as compared to rice straw (1.45 per cent) in crop residues (Table 7.1). Among the commercial feeds, highest EE content was in *Chaata* (6.03 per cent) and lowest in UMMB (1.14 per cent). There was no significant difference between different feed samples in EE content in different groups (Tables 7.2 to 7.8).

As the quality of fats or lipids, often measured in terms of the amounts of unsaturated fatty acids present in a feed, *viz.*, linoleic, linolenic and arachidonic acid, does not hold any value for ruminant animals, as these essential fatty acids are synthesized by the microbes present in the rumen. Considerable hydrogenation of the unsaturated fatty acids also does take place in the rumen. Therefore, body fats of ruminants hardly resemble with those present in the feed.

Crude Fibre

The crude fibre (CF) content of a fodder comprises cell wall contents cellulose, hemicelluloses and lignin. CF content is generally inversely related with CP value and hence is a negative attribute to the nutritive value of a fodder.

The results presented in Table 7.2 reveal that the CF content was significantly different (P<0.01) among the groups. CF range varied from 4.87 per cent (UMMB) to 48.89 per cent (finger millet straw) in different groups (Table 7.1). In tree leaves, the CF content varied from 9.89 per cent (*Prunus cerasoides*) to 23.04 per cent (*Machilus duthiei*). In grasses, CF content ranged from 26.41 per cent (mixed green grass) to 34.15 per cent (*Themeda anathera*). In crop residues, the CF content varied from 45.60 per cent (rice straw) to 48.89 per cent (finger millet straw) and in commercial feeds the CF content varied from 4.87 per cent (UMMB) to 21.69 per cent (CFB). CF content was significantly different between G1 and G2, G1 and G3, G2 and G3, G2 and G4, G3 and G4 (P<0.01), G1 and G4 (P<0.05) (Tables 7.2 to 7.8).

The CF figures in few cases are comparable with the figures found in literature (Singh *et al.*, 1988; Yadava *et al.*, 2003). The present experiment includes many fodder sources which were not chemically anlysed earlier to determine nutritive value. Hence all figures cannot be compared with the values published earlier.

Total Ash

Total ash (TA) gives an account of all major and trace elements present in a fodder sample. The major elements in a fodder include Ca, P, Mg, Na, K, S and Cl. The trace elements include Fe, Cu, Co, I, Zn, Mn, Mo, Se, Cr. These are the essential elements needed for the body structure and metabolic activities by all animals. However, non-essential elements, like Si, Ni, Ti, Al, V, B, Pb, and Sn might also be found in plants and they would also be present in the TA fraction of the chemical composition of fodders. Estimation of TA is important for suggesting nutritive value of a fodder. The higher value of essential major and trace elements is indicative of higher nutritive value.

Table 7.2: Nutritive Values (%) Comparison Between Different Tree Leaves, Grasses Fodders and Commercial Feeds

Parameters	*Tree Leaves (G1)*	*Grasses (G2)*	*Crop Residues (G3)*	*Commercial Feeds (G4)*
Dry matter**	36.07 ± 2.880	36.24 ± 12.443	89.30 ± 0.599	87.13 ± 3.622
Crude protein**	16.64 ± 1.034	6.32 ± 0.437	3.14 ± 0.185	18.32 ± 1.912
Ether extract	1.89 ± 0.248	2.13 ± 0.229	1.50 ± 0.549	3.06 ± 1.506
Crude fibre**	17.58 ± 1.008	29.30 ± 1.448	47.24 ± 0.164	10.57 ± 5.557
Total ash	12.79 ± 1.214	9.29 ± 1.194	10.31 ± 0.039	14.14 ± 5.639
Acid insoluble ash	2.74 ± 0.616	4.72 ± 0.632	5.330 ± 0.730	5.13 ± 2.103
Neutral detergent fibre**	54.03 ± 0.868	55.83 ± 1.317	68.57 ± 1.550	31.11± 13.500
Acid detergent fibre**	37.55 ± 1.500	39.98 ± 2.770	43.46 ± 0.609	18.67 ± 7.961
Digestibility**	56.76 ± 2.353	47.40 ± 2.694	52.00 ± 0.000	81.00 ± 6.658
Nitrogen free extract	51.15 ± 1.96	52.94 ± 0.774	37.25 ±1.844	53.98 ± 8.509
Organic matter	87.20 ± 1.21	90.70 ± 1.194	89.69 ± 3.624	85.85 ± 5.639
Hemicelluloses	16.47 ± 1.31	15.85 ± 3.015	25.11 ± 2.160	12.44 ± 5.539

**Significant difference $P<0.01$.

Table 7.3: Comparison of Nutritive Values (%) Between Tree Leaves and Grasses

Parameters	*Tree Leaves*	*Grasses*	*Cd 5%*	*Cd 1%*
Dry matter	$36.07^{a} \pm 2.880$	$36.24^{a} \pm 12.443$	16.89	23.09
Crude protein	$16.64^{b} \pm 1.034$	$6.32^{a} \pm 0.437$	3.50*	4.79**
Ether extract	$1.89^{a} \pm 0.248$	$2.13^{a} \pm 0.229$	1.24	1.70
Crude fibre	$17.58^{a} \pm 1.008$	$29.30^{b} \pm 1.448$	4.99*	6.83**
Total ash	$12.79^{a} \pm 1.214$	$9.29^{a} \pm 1.194$	5.35	7.32
Acid insoluble ash	$2.74^{a} \pm 0.616$	$4.72^{a} \pm 0.632$	2.46	3.36
Neutral detergent fibre	$54.03^{a} \pm 0.868$	$55.83^{a} \pm 1.317$	8.93	12.21
Acid detergent fibre	$37.55^{a} \pm 1.500$	$39.98^{a} \pm 2.770$	7.52	10.27
Digestibility	$56.76^{b} \pm 2.353$	$47.40^{a} \pm 2.694$	9.02*	12.33
Nitrogen free extract	$51.15^{a} \pm 1.96$	$52.94^{a} \pm 0.774$	8.20	11.21
Organic matter	$87.20^{a} \pm 1.21$	$90.70^{a} \pm 1.194$	5.35	7.32
Hemicelluloses	$16.47^{a} \pm 1.31$	$15.85^{a} \pm 3.015$	6.41	8.76

Values bearing different superscripts in a row differ significantly; $^{*}P<0.05$, $^{**}P<0.01$.

Table 7.4: Comparison of Nutritive Values (%) Between Tree Leaves and Crop Residues

Parameters	*Tree Leaves*	*Crop Residues*	*Cd 5%*	*Cd 1%*
Dry matter	$36.07^{a} \pm 2.880$	$89.30^{b} \pm 0.599$	24.38*	33.33**
Crude protein	$16.64^{b} \pm 1.034$	$3.14^{a} \pm 0.185$	5.06*	6.92**
Ether extract	$1.89^{a} \pm 0.248$	$1.50^{a} \pm 0.549$	1.79	1.70
Crude fibre	$17.58^{a} \pm 1.008$	$47.24^{b} \pm 0.164$	7.21*	9.85**
Total ash	$12.79^{a} \pm 1.214$	$10.31^{a} \pm 0.039$	7.73	10.56
Acid insoluble ash	$2.74^{a} \pm 0.616$	$5.330^{a} \pm 0.730$	3.55	4.85
Neutral detergent fibre	$54.03^{a} \pm 0.868$	$68.57^{b} \pm 1.550$	12.89*	17.62
Acid detergent fibre	$37.55^{a} \pm 1.500$	$43.46^{a} \pm 0.609$	10.85	14.83
Digestibility	$56.76^{a} \pm 2.353$	$52.00^{a} \pm 0.000$	13.02	17.80
Nitrogen free extract	$51.15^{b} \pm 1.96$	$37.25^{a} \pm 1.844$	11.84*	16.18
Organic matter	$87.20^{a} \pm 1.21$	$89.69^{a} \pm 3.624$	7.73	10.56
Hemicelluloses	$16.47^{a} \pm 1.31$	$25.11^{b} \pm 2.160$	9.25	12.65

Values bearing different superscripts in a row differ significantly; *$P<0.05$, **$P<0.01$.

Table 7.5: Comparison of Nutritive Values (%) Between Tree Leaves and Commercial Feeds

Parameters	*Tree Leaves*	*Commercial Feeds*	*Cd 5%*	*Cd 1%*
Dry matter	$36.07^{a} \pm 2.880$	$87.13^{b} \pm 3.622$	20.56*	28.10**
Crude protein	$16.64^{a} \pm 1.034$	$18.32^{a} \pm 1.912$	4.27	5.83
Ether extract	$1.89^{a} \pm 0.248$	$3.06^{a} \pm 1.506$	1.51	2.07
Crude fibre	$17.58^{b} \pm 1.008$	$10.57^{a} \pm 5.557$	6.08*	8.31
Total ash	$12.79^{a} \pm 1.214$	$14.14^{a} \pm 5.639$	6.51	8.91
Acid insoluble ash	$2.74^{a} \pm 0.616$	$5.13^{a} \pm 2.103$	2.99	4.09
Neutral detergent fibre	$54.03^{b} \pm 0.868$	$31.11^{a} \pm 13.500$	10.87*	14.86**
Acid detergent fibre	$37.55^{b} \pm 1.500$	$18.67^{a} \pm 7.961$	9.15*	12.51**
Digestibility	$56.76^{a} \pm 2.353$	$81.00^{b} \pm 6.658$	10.98*	15.01**
Nitrogen free extract	$51.15^{a} \pm 1.96$	$53.98^{a} \pm 8.509$	9.98	13.64
Organic matter	$87.20^{a} \pm 1.21$	$85.85^{a} \pm 5.639$	6.51	8.91
Hemicelluloses	$16.47^{a} \pm 1.31$	$12.44^{a} \pm 5.539$	7.80	10.66

Values bearing different superscripts in a row differ significantly; *$P<0.05$, **$P<0.01$.

Table 7.6: Comparison of Nutritive Values (%) Between Grasses and Crop Residues

Parameters	*Grasses*	*Crop Residues*	*Cd 5%*	*Cd 1%*
Dry matter	$36.24^{a} \pm 12.443$	$89.30^{b} \pm 0.599$	26.86*	36.71**
Crude protein	$6.32^{a} \pm 0.437$	$3.14^{a} \pm 0.185$	5.57	7.62
Ether extract	$2.13^{a} \pm 0.229$	$1.50^{a} \pm 0.549$	1.98	2.70
Crude fibre	$29.30^{a} \pm 1.448$	$47.24^{b} \pm 0.164$	7.94*	10.85**
Total ash	$9.29^{a} \pm 1.194$	$10.31^{a} \pm 0.039$	8.51	11.63
Acid insoluble ash	$4.72^{a} \pm 0.632$	$5.330^{a} \pm 0.730$	3.91	5.34
Neutral detergent fibre	$55.83^{a} \pm 1.317$	$68.57^{a} \pm 1.550$	14.20	19.41
Acid detergent fibre	$39.98^{a} \pm 2.770$	$43.46^{a} \pm 0.609$	11.95	16.34
Digestibility	$47.40^{a} \pm 2.694$	$52.00^{a} \pm 0.000$	14.34	19.61
Nitrogen free extract	$52.94^{b} \pm 0.774$	$37.25^{a} \pm 1.844$	13.04*	17.82
Organic matter	$90.70^{a} \pm 1.194$	$89.69^{a} \pm 3.624$	8.51	11.63
Hemicelluloses	$15.85^{a} \pm 3.015$	$25.11^{a} \pm 2.160$	10.19	13.93

Values bearing different superscripts in a row differ significantly; $^{*}P<0.05$, $^{**}P<0.01$.

Table 7.7: Comparison of Nutritive Value (%) Between Grasses and Commercial Feeds

Parameters	*Grasses*	*Commercial Feeds*	*Cd 5%*	*Cd 1%*
Dry matter	$36.24^{a} \pm 12.443$	$87.13^{b} \pm 3.622$	23.44*	32.04**
Crude protein	$6.32^{a} \pm 0.437$	$18.32^{b} \pm 1.912$	4.86*	6.65**
Ether extract	$2.13^{a} \pm 0.229$	$3.06^{a} \pm 1.506$	1.72	2.36
Crude fibre	$29.30^{b} \pm 1.448$	$10.57^{a} \pm 5.557$	6.93*	9.47**
Total ash	$9.29^{a} \pm 1.194$	$14.14^{a} \pm 5.639$	7.43	10.15
Acid insoluble ash	$4.72^{a} \pm 0.632$	$5.13^{a} \pm 2.103$	3.41	4.66
Neutral detergent fibre	$55.83^{b} \pm 1.317$	$31.11^{a} \pm 13.500$	12.40*	16.95**
Acid detergent fibre	$39.98^{b} \pm 2.770$	$18.67^{a} \pm 7.961$	10.43*	14.26**
Digestibility	$47.40^{a} \pm 2.694$	$81.00^{b} \pm 6.658$	12.52*	17.11**
Nitrogen free extract	$52.94^{a} \pm 0.774$	$53.98^{a} \pm 8.509$	11.38	15.55
Organic matter	$90.70^{a} \pm 1.194$	$85.85^{a} \pm 5.639$	7.43	10.15
Hemicelluloses	$15.85^{a} \pm 3.015$	$12.44^{a} \pm 5.539$	8.90	12.16

Values bearing different superscripts in a row differ significantly; $^{*}P<0.05$, $^{**}P<0.01$.

Table 7.8: Comparison of Nutritive Values (%) Between Crop Residues and Commercial Feeds

Parameters	*Crop Residues*	*Commercial Feeds*	*Cd 5%*	*Cd 1%*
Dry matter	$89.30^{a} \pm 0.599$	$87.13^{a} \pm 3.622$	29.30	40.05
Crude protein	$3.14^{a} \pm 0.185$	$18.32^{b} \pm 1.912$	6.08*	8.31**
Ether extract	$1.50^{a} \pm 0.549$	$3.06^{a} \pm 1.506$	2.16	2.95
Crude fibre	$47.24^{b} \pm 0.164$	$10.57^{a} \pm 5.557$	8.66*	11.84**
Total ash	$10.31^{a} \pm 0.039$	$14.14^{a} \pm 5.639$	9.29	12.69
Acid insoluble ash	$5.330^{a} \pm 0.730$	$5.13^{a} \pm 2.103$	4.26	5.83
Neutral detergent fibre	$68.57^{b} \pm 1.550$	$31.11^{a} \pm 13.500$	15.50*	21.18**
Acid detergent fibre	$43.46^{b} \pm 0.609$	$18.67^{a} \pm 7.961$	13.04*	17.83**
Digestibility	$52.00^{a} + 0.000$	$81.00^{b} \pm 6.658$	15.65*	21.30**
Nitrogen free extract	$37.25^{a} \pm 1.844$	$53.98^{b} \pm 8.509$	14.23*	19.44
Organic matter	$89.69^{a} \pm 3.624$	$85.85^{a} \pm 5.639$	9.29	12.69
Hemicelluloses	$25.11^{b} \pm 2.160$	$12.44^{a} \pm 5.539$	11.12*	15.20

Values bearing different superscripts in a row differ significantly; $^{*}P<0.05$, $^{**}P<0.01$.

TA content was not significantly different among the groups (Tables 7.2 to 7.8). The range of ash content in different groups was from 2.89 per cent (*Chaata*) to 20.44 per cent (UMMB). Among the tree leaves, the highest ash content was in *Ficus roxburghii* (17.33 per cent). It was the lowest in *Chaata* (2.89 per cent). In grasses, ash content varied between 6.06 per cent (*Themeda anathera* green) to 12.52 per cent (mixed green grass), in crop residues, the ash content was 10.27 per cent (finger millet straw) and 10.27 per cent (rice straw) and in commercial feeds the ash content varied between 2.89 per cent (*Chaata*) to 20.44 per cent (UMMB) (Table 7.1).

Singh *et al.* (1995) have reviewed the nutritive values of forest-based fodder species. Although most of the species analysed in this experiment were not present in their review paper, the ones that were there could be compared in terms of mineral content in some of the cases.

Acid Insoluble Ash

Acid insoluble ash (AIA) includes only silica. Silica is a negative attribute for the nutritive value of a fodder. Among the tree leaves, the highest AIA content was in *Prunus cerasoides* (6.51 per cent) and lowest in *Machilus duthiei* (0.33 per cent). In grasses, the AIA content was highest in mixed green grass (6.06 per cent) and lowest (2.35 per cent) in *Themeda anathera* green. In crop residues, the AIA content was 4.60 per cent (finger millet straw) and 6.06 per cent (rice straw) and in commercial feeds the AIA content was 0.93 per cent in *Chaata* and 7.43 per cent in UMMB (Table 7.1). The average AIA values for tree leaves, grasses, crop residues and commercial feeds were 2.74 ± 0.616, 4.72 ± 0.632, 5.330 ± 0.730 and 5.13 ± 2.103 per cent , respectively (Table 7.2). AIA content was not significantly different between the groups (Tables 7.2 to 7.8).

Neutral Detergent Fibre

Neutral detergent fibre (NDF) gives the values of cell wall contents, *viz.*, lignin, cellulose and hemicelluloses. The average NDF content of various tree species, grasses, crop residues and commercial feeds was 54.03±0.86, 55.83±1.317, 68.57±1.55, 31.11±13.50 per cent respectively (Table 7.1) and all were significantly different with each other ($P<0.01$) (Table 7.2). The NDF content varied from 46.76 per cent (*Grewia optiva*) to 59.70 per cent (*Machilus duthiei*) in tree leaves, from 52.64 per cent (mixed green grass) to 59.64 per cent (*Themeda anathera* green) in grasses, from 67.02 per cent (finger millet straw) to 70.12 per cent (rice straw) in crop residues and from 14.60 per cent (*Chaata*) to 57.87 per cent (CFB) in commercial feeds. There was significant difference between G1 and G3 ($P<0.05$), G1 and G4, G2 and G4, G3 and G4 ($P<0.01$) but not significant difference between G1 and G2, G2 and G3 (Tables 7.2 to 7.8).

Acid Detergent Fibre

Acid detergent fibre includes cellulose, lignin, lignified N, and silica. ADF determination gives an idea of the digestibility of a particular fodder. Acid detergent lignin (ADL) includes only lignin. Inextricably associated with cell wall, lignin generally is impervious to microbial enzyme action in the rumen of ruminants. Determination of lignin in a fodder is essential to establish its digestibility. ADF ranged from 27.78 per cent (*Grewia optiva*) to 43.45 per cent (*Bauhinia vahlii*) in tree leaves, 34.21 per cent (*Themeda anathera* green) to 46.82 per cent (mixed green grass) in grasses, 42.85 per cent (rice straw) to 44.07 per cent (finger millet straw) in crop residues and 9.10 per cent (*Chaata*) to 34.48 per cent (CFB) in commercial feeds (Table 7.1). ADF content was significantly different between the groups ($P<0.01$) (Table 7.2).

Nitrogen Free Extractives

The nitrogen-free extractives (NFE) include cellulose, hemicelluloses, lignin, sugars, fructans, starch, pectins, organic acids, resins, tannins, pigments, and water-soluble vitamins. There was non-significant difference amongst the groups in NFE content (Table 7.2). The range of NFE content was from 35.68 per cent (finger millet straw) to 69.72 per cent (*Chaata*) (Table 7.1). The average NFE content was 51.15±1.96, 52.94±0.77, 37.50±1.84 and 53.98±8.50 per cent in tree leaves, grasses, crop residues and commercial feeds, respectively (Table 7.2). There was significant difference ($P<0.05$) between G1 and G3, G2 and G3, G3 and G4 (Tables 7.3 to 7.8).

Organic Matter

The average OM content was 87.20±1.21 per cent in tree leaves, 90.70±1.19 per cent in grasses, 89.69±3.62 per cent in crop residues and 85.85±5.63 per cent in commercial feeds. There was no significant difference in OM content in different groups (Table 7.2). Highest OM content was 97.11 per cent (*Chaata*) and lowest 80.90 per cent (CFB) among the groups (Table 7.1).

The OM values of some of the fodder species were comparable with those reported by Yadava *et al.* (2003). Many of the species analysed chemically in this study did not appear in earlier reports and hence they cannot be compared.

Hemicelluloses

There was non significant difference between the groups in hemicelluloses content (Table 7.2). Highest hemicelluloses (HC) content was 25.43 per cent in *Themeda anathera* green and lowest 5.5 per cent in *Chaata* (Table 7.1). The average hemicelluloses content was 16.47±1.31, 15.85±3.01, 25.11±2.16 and 12.44±5.53 per cent in tree leaves, grasses, crop residues and commercial feeds, respectively (Table 7.1). There was significant difference ($P<0.05$) between G1 and G3, G3 and G4 (Tables 7.2 to 7.8).

Dry Matter Digestibility

Nylon bag dry matter digestibility (DMD) was significantly different ($P<0.01$) among the groups (Table 7.2). The digestibility values ranged from 40.00 per cent (hay of *Themeda anathera*) to 90.00 per cent (*Chaata*) (Table 7.1). The average digestibility was 56.76±2.35 per cent in tree leaves, 47.40±2.69 per cent in grasses, 52.00±0.00 per cent in crop residues and 81.00± 6.65 per cent in commercial feeds (Table 7.1). Digestibility was highest (67.00 per cent) in *Grewia optiva* and lowest (40.00 per cent) in *Ougeinia dalbergioides* in tree leaves. In grasses, it was the highest (55.00 per cent) in *Themeda anathera* green and the lowest (40.00 per cent) in the hay of *Themeda anathera*. In crop residues, it was the highest and the lowest in finger millet straw and rice straw, respectively. In the commercial feeds, the highest value (90.00 per cent) was recorded in *Chaata* and the lowest (68.00 per cent) in CFB (Table 7.1). Digestibility was significantly different in G1 and G2 ($P<0.05$), G1 and G4, G2 and G4, G3 and G4 ($P<0.01$) but not between G2 and G3, G1 and G3 (Tables 7.2 to 7.8).

Many a workers (Singh 1985; Singh and Naik 1987; Singh *et al.*, 1988, 1990; Singh *et al.*, 1995; Bohra 2003; Yadava *et al.*, 2003) have already attempted to determine nutritive and digestibility values of uncultivated forest–and agroforestry-based fodders. While in most of the cases our values are in conformity with theirs, we have included some more fodder species in our analyses and the values presented in this analysis perhaps for the first time give some more important clues about the potentials of rangelands serving as the appropriate habitats for fodder species.

Chapter 8
Rangelands, Ecological Integrity and Sustainability

Ecological integrity of a system, *e.g.*, a mountain agro-ecosystem, is a precondition to sustainability. Rangelands that occupy chunk of the geographical area in the Himalayan mountains undoubtedly have a key role to play towards imparting high degree of sustainability to the agro-ecosystem and thereby to the land-based livelihoods in the mountain areas. Ecological integrity of a system is ensured when there are viable ecological linkages between components of the system and, as a result, there is considerable flow of nutrients within the system. Factors influencing bio-geochemical cycles, water cycle and climate regulation are pivotal for ensuring ecological integrity at macro-level of the geographical area. The uncultivated rangelands covered largely with natural vegetation are pivotal towards generating, regulating, and optimizing factors that infuse ecological integrity into the whole system.

Rangeland-Livestock-Farming Linkages

Fragility is one of the inherent characteristics of Himalayan mountains. Problems linked to fragility are aggravated as a result of large-scale damage to the environment which, in addition to deforestation, is also contributed by continuous ploughing and cultivation of food grain crops. Rangelands maintained in the agro-ecosystems are believed to be an appropriate response to the natural fragility.

In the mountains, area under cultivation is very small, which is also attributable to the maintenance of rangelands in the agro-ecosystems. Had the farming communities gone for reckless terracing of mountain slopes at the expense of uncultivated forest / grassland areas, the current area under cultivation would have been pretty large and agrarian picture in the mountain areas would have been altogether different. Problems of the fragile ecosystems then would have been many times more aggravated, perhaps uncontrollable. The farming systems involving large areas maintained under rangelands, as of today, have been evolved by the local farmers through trial and error over millennia.

Traditional agriculture and animal husbandry in the mountains establishes the vital rangelands-livestock-cropland linkages. Livestock feed on range plants and retain a proportion of energy and nutrients for their maintenance, while the rest is converted into draught power and products (milk, wool, meat, etc.). A greater proportion is voided as dung and urine which, as manure, are transferred to the cropland for the maintenance of soil fertility inevitable for food production. Fodder biomass from croplands is fed to livestock and a proportion of energy and nutrients, as manure, is recycled into the same soil. Mediated by livestock, this nutrient flow contributes to the essential ecological integrity and sustainability of mountain agro-ecosystems. This has been discussed elsewhere in the text.

Rangeland Utilisation for Life-support Activities

Rangeland ecosystems are utilised for a number of reasons vital for livelihood systems in the mountain areas. The villagers–*i.e.*, the livestock-depended farming community–have some preferences for the utilization of rangelands. On the basis of personal interview with the folk population, people's preference ranking has been presented in Table 8.1. The actual number of species local farmers can readily identify and quickly enumerate is far greater than shown in the table. The average number would also vary from place to place in the mountains.

Table 8.1: People's Preference of Rangeland Utilisation

Use	*Preference Ranking*	*Species No.**
Grazing	1	All palatable spp.
Fodder lopping	2	15
Fuel wood	3	Virtually all available woody species
Agricultural tools and implements	5	8
Medicine	9	19
Dyes	10	3
Fruits	11	15
Other edible products	6	17
Timber	8	6
Bedding material	4	6
Mulch	7	6
Non-consumptive use value	Over and above	All available spp.

* Species number is according to what farmers can identify and enumerate at a glance; all the species, except for grazing, include only trees and shrubs; the real value is considerably higher and would change from area to area and from place to place in the same area.

Rangelands' first and the foremost utilization is for grazing the livestock. Keeping livestock on grazing during daytime is one of the traditional and cost-effective ways of livestock production in the Uttarakhand Himalaya as well as elsewhere. Tribal populations occupying upper reaches of the Himalaya owe very large herds especially of goats and sheep and utilize rangelands for livestock production more effectively. Nomads like Gujjars occupy very large number of buffaloes which graze in the alpine meadows during summer months and the lower ranges during winter. All types of land uses and virtually all palatable species available in the rangeland where livestock are to graze are utilized by livestock during the process of grazing.

Fodder lopping is number two economic activity. All the tree fodder species available in the rangelands in an area are utilized for fodder lopping by the livestock owners. In Uttarakhand rangelands about 550 species of fodder plants are utilized during different seasons of the year. Farmers can enumerate some 15 types of fodder plants, especially trees, occurring in their immediate environments.

Collection of bedding material to be used in the livestock sheds comes at number 4. There are six species people generally use for bedding material. The bedding material mixed with dung and converted into manure is eventually used for manuring the crop fields.

Agricultural tools and implements are made of certain woody species occurring in the rangelands, which is the fifth preferable use of these ecosystems in the mountains. Eight species are generally used for making agricultural tools and implements.

Rangeland ecosystems are sources of several edible products. Collecting these products for consumption purpose is the sixth preference of the people. At number seven comes the use of rangeland-based species for mulch.

Extraction of timber is the eighth preferable economic activity of the people. Medicine, dyes and edible fruits come next in order in that sequence for which people depend on rangelands.

These uncultivated areas, in fact, have potential to provide more quality food per unit area on sustained basis than the cultivated land. People in the mountains have been extracting food (edible fruits, flowers, buds, seeds, vegetables, mushrooms, honey, etc.) from the forests/ rangelands for centuries. Rangelands are often appreciated for their contribution to the community of fuel, fodder, and timber. That they can provide the most basic necessity of humans, *i.e.*, a variety of food items, is seldom appreciated. Singh (2004a, 2004b) has brought to the fore vital role of uncultivated forest areas for providing different food products to the community which is evident from his study in the Henwal Valley of Garhwal in Uttarakhand Mountains. In this Valley, as many as 189 different naturally occurring species of plants offer edible products while people cultivate only 92 different food-producing plants on their cultivated land. Further, uncultivated foods are more nutritive than the cultivated ones and hardly require use of fuel energy for processing. Comparing this scenario with that of the plains' agriculture in the adjoining areas of the Garhwal mountains, we would find that farmers depend only on 33 plant species for their survival. Thus, from the point of view of plant species diversity offering variety of nutritive foods, rangelands are the richest sources.

Apart from several use values, there is also non-use values associated with rangeland ecosystems as indicated in Figure 8.1. These values are over and above the direct use values. Banskota (2004) and Shiva *et al.* (2005) have illustrated the use and non-use values of several well-known and lesser known species in the mountain areas.

Nutrient Flows

Nutrients are a collection of chemical compounds, minerals and elements essential to the survival of living organisms (Bourn *et al.*, 2005). Plants take up the nutrients from the soil reserves and atmosphere and accumulate in their bodies. Animals take up these from the plant biomass they consume. Not all the nutrients are deposited in the bodies of plants and animals, but certain proportion of these flows through the environment. This passage or flow of nutrients in the environment is a nutrient cycle. In a Himalayan agro-ecosystem, livestock play vital role in the nutrient cycling.

Nutrient cycles are a sub-set of broader class of global biogeochemical cycles, including water, carbon, oxygen, nitrogen and mineral cycles (Bourn *et al.*, 2005). Biogeochemical cycles–especially those of NPK–are significantly influenced by agriculture, the single anthropogenic activity responsible for major ecosystem transformations the world over. Livestock have to play a very crucial role in the recycling of nutrients by voiding dung and urination while grazing. Manure collected in the livestock sheds is also transferred into crop fields by humans and thus the nutrients are recycled into soils. The nutrients disposed into cropland soil are not entirely to be obtainable through crop residues. Major chunk of the nutrients comes from uncultivated rangelands which are transferred to cropland soils. Since grazing length in mountain areas under sedentary system of livestock rearing is relatively shorter and not all animals (*e.g.*, buffalo species) are reared through daytime grazing, nutrients to be recycled into rangelands are relatively less in amount. The rangelands, particularly the upland forests, by and large, are

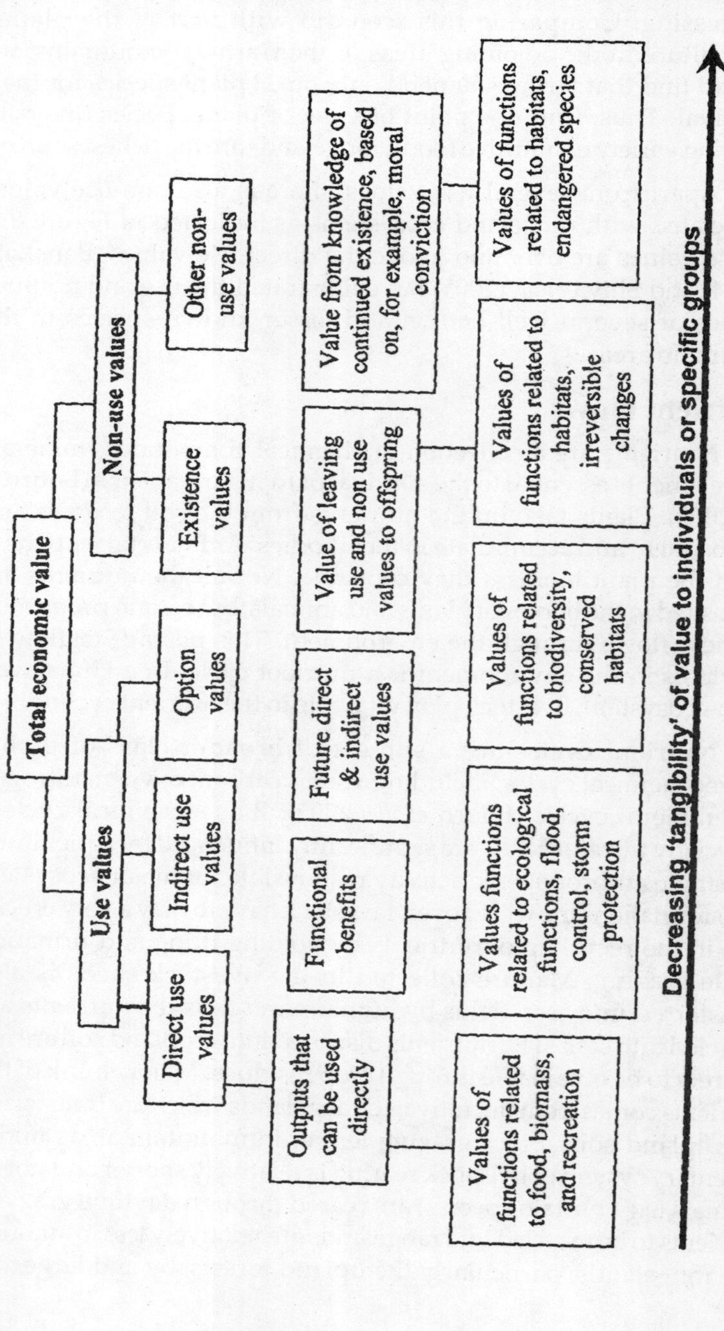

Figure 8.1: The Economic Value of Rangeland Biodiversity (*Source*: adapted from Banskota 2004)

the independent solar-powered ecosystems and hence they are ecologically more stable than the croplands which are more fragile and thrive thanks to subsidy received in the form of nutrients from the uncultivated areas, particularly via livestock. Livestock-mediated flow of nutrients, therefore, is critical to the productivity of the ecologically vulnerable croplands/ cultivated areas and not to the ecologically more stable rangeland ecosystems. Deep-rooted trees and a variety of other woody perennials and herbaceous vegetation in the uncultivated ecosystems take care of the natural fertility maintenance of the agro-ecosystem. Greater degree of ecological degradation, however, adversely affects the pace of nutrient flow from the rangelands to the cultivated areas. Optimum amount of nutrients to flow into croplands is a precondition to the fertility management of the ecologically more fragile and more vulnerable system.

In mountain agro-ecosystems, such as in Uttarakhand and other parts of the HKH Region, there is no considerable import of nutrients from outside the region. All the nutrients are produced from within the system itself. Moreover, livestock are not dependent on imported nutrients as would be the case in 'industrial' livestock production units.

In many countries and areas of the world where–under the industrial livestock production system–the animals are kept in high density in piggeries, poultry farms and feedlots, they are dependent entirely on cereal-based diets containing supplementary NPK and various heavy metals, including Cu and Zn. In such a system, relatively small proportions of the supplementary nutrients are absorbed, with most passing through the gut to be excreted with other animal wastes. Net imports of animal nutrients lead to local accumulations and excess soil nutrients, with adverse effect on soil fertility, increased run-off, eutrophication of aquatic systems and contamination of ground water and drinking water supplies (Bourn *et al.*, 2005).

In sharp contrast to the intensive high input/ high output systems (including dairying, feedlots, piggeries and poultry farms) as maintained in the industrialized countries are the extensive low input/ low output production systems including pastoralism and complex mixed farming systems as operating in the Himalayan mountains. The latter type of the livestock production systems is environmentally sound and more sustainable. Animal-nature interactions in Himalayan mountains' smallholder systems indeed are environment-friendly. Livestock are the most important dynamic factor in managing the ecological or organic faming systems evolved in the mountains over millennia.

Ecological Integrity and Sustainability

An ecosystem is a structural and functional unit of nature which is self-sufficient and in which there is a definite pattern of nutrient and energy flows vital for the sustenance of the three organisational levels of life, *viz.*, producers, consumers and decomposers. For that matter, agro-ecosystems in the mountains of the Himalayan Region are also self-sufficient and an independent functional unit of nature. Himalayan agro-ecosystems have rangelands, livestock and croplands as their integral components. Livestock consume the biomass growing in rangelands a part of which is used to manure croplands. Plant biomass for use as mulch and bedding material extracted from rangeland components also ends up as manure in the croplands. Much of the manuring of the cropland areas is thus done by the nutrients contained in the biomass produced in the rangelands.

Comprising a variety of deep-rooted trees and shrubs along with herbaceous plants and a variety of other wildlife forms, a rangeland ecosystem is very complex in its functioning and ecologically more stable than the cultivated land. A thin layer of fertile soil supports shallow-rooted (sometimes deep-rooted) food crop annuals. This soil requires to be constantly nourished for fertility maintenance. The cultivated land therefore is ecologically more fragile and vulnerable. Management of this fragility and vulnerability requires constant replenishment of the thin layer of soil in a cropland subsystem. The cropland soils are fed with the nutrients contained in the biomass produced in a rangeland ecosystem. Crop residues produced in the cropland are also fed to the livestock a part of which also ends up as manure.

Livestock constitute the dynamic component of a farming system helping nutrient flows in two ways: transfer of nutrients from ecologically complex and relatively more stable rangelands to the more fragile croplands and recycling of nutrients into the cropland. Livestock thus serve as the living agency to mediate nutrient flows in mountain agro-ecosystems. A part of the energy consumed is retained in their bodies to be used for maintenance and production.

Rangelands contribute to enhance complexity and impart considerable ecological stability to the whole agro-ecosystem. They are also managed for sustainable production of useful biomass. Rangelands provide variety of fodders for livestock feeding as well as other useful biomass, such as wild fruits, uncultivated vegetables, fiber, medicinal plants, honey, etc.

Maintenance and utilisation of rangelands in the mountains is a unique case of community-based management. Rangelands are spread over community lands (sometimes referred to as commons) to be shared by the residents of a *Gramsabha*, a cluster of some three to five villages. Each *Gramsabha* has its own rangeland, which, for the local community, is a "jungle". Utilisation of rangeland products is for the entire community of the *Gramsabhas*.

Fodder production and its utilization is the major concern of the Forest Panchayats, the forest management bodies in Uttarakhand. Since bulk of the fodder necessary for livestock production is produced in the rangelands, fodder and livestock always remain in the focus of the farming communities in the Himalaya. Rangelands through their several attributes and functions contribute to ecological integrity and, consequently, to sustainability of the whole agro-ecosystem.

Ecosystem services rendered by an ecologically sound ecosystem are more intense and more useful. These services emanate from well-managed rangelands and help maintain an environment which is congenial for the productive services of a cropland. Conservation of myriad life forms in its own boundaries, moisture circulation in the whole farming system, and maintenance of appropriate micro-climate are the other intangible attributes of the rangeland ecosystems in the mountain agro-ecosystems. The intangible ecosystem services are often not taken into consideration but are very vital for the efficient functioning and performance of the whole agro-ecosystem on which livelihood systems of rural communities are dependent. A rangeland is a solar-powered ecosystem. The upland terrestrial solar-powered ecosystems are independent (those in the lower areas often receive natural subsidies particularly in terms of nutrients).

Table 8.2: Different Attributes of Rangelands as Inputs to Ecological Integrity

Attribute	*Input to Ecological Integrity*
Biodiversity Conservation	1. Immense biodiversity of plants, animals and microorganisms varying according to altitude, slope-orientation, relief, rainfall pattern and other environmental factors;
	2. Wild animals' contribution to plant dispersal and energy and nutrient flows;
	3. Conversion of solar energy into chemical energy through photosynthesis;
	4. Contribution to ecosystem resilience and efficiency;
	5. Response to the inherent fragility of the mountains;
	6. An inherent ecosystem service vital for systemic linkages.
Soil and Water Conservation	1. Conservation of the basic natural resources helps in enriching the very base of life activities in the and in building up an environment suitable for habitat ecological regeneration;
	2. Circulation of moisture within and amongst ecosystems
	3. Contribution to maintenance of vital hydrological cycle;
	4. Life imbibes in the soil also and it functions as a wonderful ecosystem;
	5. Nutrient flows through soil and water as natural subsidy for other ecosystems in lower areas;
	6. Contribution to temperature regulation and micro–and macro climate.
Food Chain and Food Web	1. Various trophic levels imbibing with diversity of life thrive
	2. Unidirectional transfer of energy from the organisms of one trophic level to those of the other;
	3. Ecosystem complexity is increased, which, in turn, enhances sustainability
	4. Nutrient cycling through the action of decomposers;
	5. Energy and nutrient flows from rangeland to cropland via livestock contributing to sustainability of farming systems.

These forms of ecosystems–unlike the fossil fuel-powered ecosystems of the cities and industrial areas–demonstrate considerable resilience and are self-sufficient and sustainable. The cropland, on the contrary, can only be sustainable when it constantly receives natural subsidy in the form of nutrients in organic matter. Referred to as nature-subsidised solar-powered ecosystem, the cropland in a Himalayan agro-ecosystem is less resilient, more vulnerable and often attaining the very high degree of fragility.

Major attributes of rangelands to the maintenance of the indispensable ecological integrity are summarized in Table 8.2. Ecological integrity, in essence, is function of the ecosystem, such as conservation of biodiversity, soil and water and energy and nutrient flows through food chains and food web, etc. Ecosystem functions/ services are the inevitable products of ecological integrity. In the state of ecological integration, cropland and rangeland ecosystems function as a single independent functional unit, the agro-ecosystem. Livestock provide a crucial dimension to this integrity. In the state of isolation, the cultivated land would require inputs from the market system, as would be revealed in case of the food production systems under the umbrella of Green Revolution in favorable areas in the plains. Sustainability, in essence, is the function of this ecological integrity.

Chapter 9
Traditional Management of Range Resources

Animal husbandry is older than crop cultivation. Human civilization, in fact, entered into a new phase of its development and evolved to the present state thanks to the adoption and practices of animal husbandry. Range management appears to be as old as animal husbandry. And so is the knowledge relating to range management. Range management also appears to be the oldest art of natural resource management. Since the dawn of civilization, natural resource management systems have been the local community-based systems. Indigenous perceptions and practices regarding management of natural resources primarily concerned with protection and utilisation of biologically diverse resources. Human-managed natural resource systems endowed with nature's biodiversity virtually are as diverse as the human-inhabited ecosystems on Earth.

The Himalayan region has unique functions and roles, and range management by local farmers and transhumant societies is the largest intervention into it. Throughout the history the interaction between mountain people and the natural systems had helped communities to maintain the richness of species, communities and genetic materials in both the productive systems and wild lands of the mountain environments (Shengji 1994).

There have been three fundamental means to manage rangelands in the Himalayan mountains, particularly in its north-western region: i) burning of the forests; ii) cutting down and/ or thinning of the forests; and iii) abandonment of cultivated lands.

Amongst these, the former has been the method that could be easily handled by people. Slash-and-burn farming, also called the swidden farming, shifting farming or *jhoom* cultivation in the north-eastern hills, is different from that practiced in the north-western Himalayan mountains. While in the former case the cleared forest area is used for cultivating food and/or cash crops and place of agricultural activities is changed after a cycle of few years, in the latter, the forest area is managed for pastoral activities and it is a part of the settled agriculture.

Cutting down or thinning of a forest area is an old art of silvi-pasture management. It involves an appropriate mix of woody species and ground flora of herbaceous plants dominated by grasses. Such a management of specific vegetation involving economic species to provide specific products (foods, fodder, fuel wood, fibre, timber, medicines, etc.) and specific functions (soil and water conservation, micro-climate maintenance, synergy, etc.) in the rangeland ecosystems is a wonderful articulation of local farmers' traditional knowledge.

Burning and cutting down and/or thinning of forests allows the herbaceous species (graminae and non-graminae families) to thrive on ground exposed to more solar radiation. These species are of special socio-economic importance. In a rangeland, the first preference is given to the fodder species. The ground flora is virtually used as fodder for livestock. However, woody species occurring in a rangeland ecosystem, through their leaf biomass, also contribute significant amount of fodder to farmers for their livestock. All woody species are often used for extracting fuel wood while many others may be of multiple economic uses simultaneously providing other useful products such as fruits, edible seeds, dyes, medicine, fibre, bedding material, minor timber, agricultural implements and tools, etc.). Some species may be of little direct economic importance but are maintained to render useful ecosystem services vital for the maintenance of agricultural and livelihood systems in mountain areas.

Abandoned agriculture on a piece of land also results in the development of a rangeland ecosystem. This land is often predominated by grasses and other herbaceous plants interspersed with shrubs and trees, which were originally kept as part of agro-forestry system. The trees and shrubs are broadleaved extensively used for top lopping.

There is yet another practice of farmers in which cultivated land is used as grazing ground during certain period of the year. This is the fallow practice. Terraced fields used for cultivation are left fallow for a crop season in a year. During this season, livestock can graze on the fallow fields. Grasses of field bunds, green plants, or weeds, growing in the absence of ploughing, leftovers of the previous food crop (leaves, stems and often roots) are grazed by the animals. The private cropland left fallow is also used as CPR during the fallow period. Fellow practice leads to increase in grazing area temporarily. This is the practice adopted by farmers for centuries to give periodic rest to their fields for maintaining fertility of the soils. The soil fertility through fallowing practice is managed by i) preventing draining out of nutrients via food crops, ii) allowing microbial degradation of crop residues (mostly roots), iii) *in-situ* manuring by the grazing animals, and iv) green manuring upon mixing up green plants/weeds through ploughing at the end of fallow period.

Indigenous knowledge (IK) is the knowledge of the people living in a certain area, generated by their own and their ancestors' experience and including knowledge originating from elsewhere but has been practically accepted by the local people. It is also referred to as local knowledge (Reijntjes *et al.*, 1992).

Rural people of the Uttarakhand Himalaya are rich in the knowledge about the native plants, most of which are occurring in uncultivated (forest/ rangeland) areas. Rare examples relating to Himalayan ethnobotany were highlighted by Gaur (1999, 2004), Gaur and Tiwari (1987) and Gaur *et al.* (1980, 1984, 1988, 1992) indicating the prevalence of indigenous knowledge in the study area.

Site-specific Management

The natural resources, techniques and strategies from which farmers can choose in creating, maintaining and developing their farm systems involving rangelands are largely determined by the specific characteristics of the ecosystems. Local farmers have normally investigated local options to the greatest possible extent and exploit them well. In this way, finely tuned farming systems have developed which have survived for generations. The local farming community relies on physical and biological diversity. Advantage is taken of complementarity between crops, animals and pastures. Each farming community has developed different natural resource/ rangeland management techniques that suit the specific local conditions (Reijntjes *et al.*, 1992).

Local farmers used to manage a blend of diversity of trees, shrubs, and herbaceous plants by conserving, utilizing, planting, and protecting in accordance with the site specificities. They are well acquainted with a particular site and the type of vegetation that can grow over there. Altitude, slope-orientation, and other topographic or physiographic factors–including edaphic factors and microclimate–are the main basis of the livestock-dependent local farming communities to manage their rangeland resources.

Integrating Range and Livestock

Available rangeland vegetation is converted into milk, meat, wool, draught power, etc., by means of rearing animals and using range vegetation as source of their feed. This helps the farmers increase subsistence security by diversifying the food generating and other economic activities. In addition, this integration also helps transfer energy nutrients between animals and cultivated crops via manure and via use of draught power. Livestock can enhance cropland productivity by intensifying energy and nutrient flows.

Rangeland-livestock integration is one of the key strategies of the livelihood systems evolved by mountain farmers. A rangeland ecosystem stocked with enormous biodiversity of trees, shrubs, herbaceous plants, wildlife, and micro forms of life is a rich store of energy and nutrients and is ecologically more stable than a cropland.

Biodiversity of rangeland ecosystems has its bearing on the biodiversity of food crops in cropland areas thanks to the range-livestock-farming integration. By means of the diversification of food production activities ensured through this integration, a farming community enhances the level of its food security.

Livestock Management

Mountain farmers rear several species of livestock to derive full advantages of the available natural resources in their environments. By maintaining diverse species–*e.g.*, cattle, buffaloes, goats, sheep, yaks, horses, mules, ponies, rabbits, poultry, etc.–farmers tend to exploit a wider range of feed resources. This might not have been possible with one or two species. Cattle prefer grasses and herbs and so do sheep. Sheep, however, would resort to browsing during the scarcity of the preferred fodder. Goats are browsing animals. They eat the leaves of shrubs and trees. Buffaloes generally prefer stall feeding, but transhumant communities, like Gujjars, manage their large number of buffaloes under grazing over vast tracts of rangelands from the foothills to alpine zone of the Himalaya. Yaks (rarely reared in Kumaun Himalaya) take full advantage of the alpine meadows adjacent to perpetual glaciers. Poultry birds can thrive on foraging around homesteads and on kitchen waste.

Transhumant pastoralists select the species and breeds of animals that are capable of walking long distances and covering treacherous and rugged terrain. The animal breeds farmers rear are suitable and well-adapted to mountain environments, can subsist on local feed resources, have high feed conversion efficiency and resistance against diseases.

Ethnovet Medicine

Conventional animal health management is exorbitantly expensive and unaffordable by small and marginal farmers who largely depend on their livestock resources for their livelihoods. The modern health cover essentially requires external veterinarians who are not well acquainted with the problems of some of the inaccessibility–and poverty-ridden areas, such as the Himalayan mountains. Ethno-veterinary practices in the Himalayan mountains areas are not only common but form a dominant system of livestock health management, as well as offer a cornucopia of benefits to the poor livestock-dependent communities as well as to the entire livestock sector. Remote, inaccessibility-ridden poor areas inhabited largely by tribal and pastoral communities are especially rich in the ethnovet-related knowledge. In the garb of modern veterinary advances, such traditional knowledge systems appear to have lost ground and hence need to be recognized, re-examined, revived and promoted. Gaur *et al.* (1992), Tripathi and Mandape (2002), Joshi and Mandape (2004), Joshi and Singh (2004), Mandape *et al.* (2004), Joshi *et al.* (2005) and Joshi and Singh (2006) have documented some ethnovet practices prevailing in the rural areas of Uttarakhand Himalaya. Gaur (1999) has given panorama of the plant diversity present in diverse ecosystems in the Garhwal district of Uttarakhand Himalaya, along with ethnobotanical notes of each species.

There are a number of medicinal plants occurring in diverse ecological niches which are of great ethnovet value and used for treatments of various ailments by local farmers. In mountain areas, *Aegle marmelos, Aloe vera* and *Trigonella foenumgraecum* plants are used for the treatment of burns. *Annona squamosa, Azadirachta indica, Chenopodium album, Eclipta alba* and *Lens culinaris* are vital for the quick recovery from fresh and cut wound. *Caryopteris odorata, Girardinia diversifolia, Grewia optiva, Murraya paniculata* and *Pinus roxsburghii* are used for the treatment of bone fractures. *Asparagus officinalis* plant is utilized in increasing milk production in lactating mothers and is also used in health tonics. *Ocimum sanctum, Melia azadirachta, Tagetus erectus, Hibiscus rosa chinensis, Phyllanthus niruri, Crinum asiaticum, Vetiveria zizanioides, Cannabis sativa, Bacopa monnieri,* etc. are other plants of ethnovet value extensively used by farmers in the mountains (Joshi *et al.*, 2005).

In the Terai and plain areas, *Triginella foenumgraecum, Trachyspermum ammi, Carcica papaya, Zinziber officinale, Allium sativum* and *Saccharum officinarum* are used for the treatment of respiratory disorders in livestock. *Euphorbia hirta, Pistia stratoites, Syzigum jambolanum, Cyamopsis psoralioides, Cuscuta reflexa* and *Cuminum cymium* are vital for curing digestive disorders in livestock. *Acacia arabica, Brassica nigra, Azadirachta indica* and *Allium sativum* are used in skin disorders. *Allium cepa* and *Tinospora cordifolia* are used for curing fever while *Ficus bengalensis* and *Cocus nuficera* are used for curing swelling neck. *Aegle marmelos* and *Ricinus communis* are used for evacuation of dead calf (Joshi *et al.*, 2005).

Tribal belts of the Greater Himalayas are especially rich in the ethno-vet practices. The environments in this area are especially suitable for an exceptionally rich variety of medicinal plants vital for the treatment and control of common and rare livestock diseases. Use of the rare herbs is made in case of pregnancy, at the time of parturition, for sustaining milk yields, improving quality of livestock products as well as for the management of specific health problems. Some herbs are common for human and livestock health management (Joshi *et al.*, 2005; Joshi and Singh 2006).

Knowledge regarding specific medicinal plants and their usage is on the tips of the livestock farmers. Farming communities inhabiting Terai and plains of the Indo-Gangetic valley have lost much of the knowledge relating to the medicinal plants and their usage in health management. Some old people, however, would readily identify the plants and their uses in specific diseases/ problems.

New generation is more dependent on modern vet medicines. Ethno-veterinary practices go hand in hand with traditional farming. Transformation of agriculture leads to the erosion of these practices. The other dominant factor is the degree of inaccessibility. Higher the degree of inaccessibility, as in the mountain areas, higher would be the dependence on ethno-medicines (Joshi *et al.*, 2005).

Ethno-vet practices of the farming class have been evolved over millennia through trials and errors. They should be regarded as futuristic rather than condemned as a remnant of the past. Recognition, documentation, demystification, augmentation, promotion and articulation of this traditional knowledge system would essentially contribute to the amelioration of livestock-based production systems and economic returns to the farmers.

Grazing Management

Ecologists today often make advocacy for deferred-cum-rotational grazing. The pastoralists in the Himalayan areas, however, have been adopting this practice for decades, probably for centuries. They do not graze their animals on a single patch in a limited area, for they are well aware that frequent defoliation eventually ends up in the cessation of plant regeneration. They keep shifting grazing area and closing the earlier grazed area for giving enough time for plant regeneration, and so on. The transhumant communities are especially conscious about changing grazing grounds according to seasons and thus about the maintenance of rangelands' health.

Browsing animals, such as goats, are useful in reducing encroachment of unwanted bushes into pasture. The selective grazing habit of animals can be deliberately used to control 'weeds'.

Fields left fallow are used as CPRs, especially for grazing animals. While grazing, animals void dung and urine and fertilise the soil. This practice is called kraaling. In Nepal and parts of the Indian Himalaya, farmers have to pay the owners of goats and sheep against the kraaling. In-situ manuring, *i.e.* the manuring of field by tethering the animals directly in the fields is an important strategy developed by the mountain farmers over ages. During dry season, cattle (and small ruminants) graze harvested fields of finger millet left fallow for fertility regeneration and manure them directly. The draught animals also directly manure the fields, in which they perform agricultural operations by voiding their dung and urine. Results of an experiment by Powell and Ikpe (1992) showed that millet yields in kraaled areas (which received both manure plus urine) were over twice as high as yields where only manure from stall-kept animals was spread by hand. Residual benefits of urine were also observed. Yields in plots where cattle and sheep were kraaled the previous year were about 1.5 times greater than where only manure had been hand-spread (Powell and Ikpe 1992).

Grass Harvesting

Vegetative propagation and sprouting of seeds are the two means of plant regeneration in the ecosystems. Regeneration by seed shedding is taken advantage of so that root reserves of energy are not exhausted by frequent cuttings. The grass is harvested after shedding of seed in the months of October-November. This is the natural way of allowing ecological regeneration of rangelands of which the farmers are well acquainted.

Grass Distribution

Managed in protected plots in the village rangelands, the grass is distributed to all families on the basis of some egalitarian principles, and social sanctions work in this regard. Harvesting is to be done by the family members, invariably by women, who have to carry the load to home. A family would bring equal numbers of bundles in a day until all the grass is harvested. Sometime all families would be allotted equal number of days for harvest and bring home as much grass as one can during those days. Sometimes this arrangement is done that a woman of a family would bring as many head loads of grass as she can during a day harvest. Sometimes numbers of animals with a family would be the criterion of distribution. Grass distribution pattern thus would depend on the mutual decision of the people of a gramsabha.

Livestock Feeding

Farmers have their own strategies of livestock feeding. They have deep knowledge about the nutritive value of the feeds and fodders they offer to their animals. *Bhimal* (*Grewia optiva*) and *kharik* (*Celtis australis*) are regarded to be the best tree-based green fodders. Their knowledge about these two tree fodders is in conformity with the published reports on the nutritive values of tree fodders of mountain areas. These two trees provide leaves with high crude protein value free from incriminating factor tannin and with high digestibility value (Singh and Naik, 1987; Singh *et al.*, 1995). They would also prefer other tree species like *Bauhinia, Ficus, Ougeinia, Vitis,* etc. which are also known to offer highly nutritive fodder.

Farmers select appropriate fodders in different seasons and try to make balance of nutrients in livestock feeding. They have all knowledge relating to the lopping cycle or fodder calendar. They respond to lean period (period of fodder scarcity) by way of incorporating leaves of evergreen trees and/ or field weeds and by nutrient supplement through concentrate feeding.

Farmers prepare their own concentrate, the feed rich in total digestible nutrients and energy, incorporating home-produced food grains. The ingredients used are black soybeans (10 per cent), soybeans (15 per cent), barnyard millet (10 per cent), gram (5 per cent), wheat (45 per cent), barley (10 per cent) and sugar (5 per cent). Locally called *chaata* in the Kumaun Himalaya, this concentrate is supplemented in the diet of pregnant and milking animals and of working bullocks.

Farmers provide dietary allowance to pregnant cows and buffaloes and to milking animals. This is provided through concentrate supplement. Dry animals (*i.e.*, animals not in milk) are seldom provided concentrate. Bullocks during idle days are also not fed concentrate. They, however, are given nutrient-rich diet during the days when they are put to work. Goats and sheep are kept exclusively on grazing.

Farmers offer specific feeds and fodders to increase quantity and quality of milk. The straws of *Echinochloa, Oryza, Hordeum, Eleusine, Triticum* and *Pennisetum* are given to livestock, which is an age old practice. There are straws of a number of cultivated plants which do not make any impact on milk production. Besides the straws of cereals and legumes, there are several plants used for fodder. These plants directly affect the quantity as well as quality of milk. Local inhabitants are well acquainted about it. Several wild plants increase the nutritional value of the milk and lopped severely for the purpose. On the other hand, there are some plant species which often decrease the quality of milk. Such types of species are used as forage during scarcity or famine period.

Gaur *et al*. (1983) and Negi *et al*. (1985) have presented a list of the plant species in Garhwal Himalaya that increase or decrease the quantity and quality of milk. Some of these plants are being named below.

Plant species that increase milk quantity and quality and identified by the farmers are: *Acacia farnesiana* (L.) Willd., *Bauhinia purpurea* L., *B. vahlii* Wight and Arn., *B. variegata* L., *Bothriochloa glabra* (Roxb.) A. Camus, *Carpinus viminea* Lindley, *Cassia tora* L., *Caesalpinia decapetala* (Roth.) Alton, *Castanopsis indica* Roxb., *C. tribuloides* (J.E. Smith) A. DC., *Celtis australis* L., *Chenopodium album* L., *C. murale* L., *Chrysopogon gryllus* (L.) Trinius, *Cissampelos pareira* L., *Clematis barbellata* Edgew, *C. grata* Wallich, *Crotolaria medicaginea* Lam., *Cyananthus lobatus* Wallich ex Benth., *Cynodon dactylon* (L.) Persoon, *Danthonia jacquemonti* Bor., *Debregeasia salicifolia* (D. Don) Rendle, *Dichanthium annulatum* (Forsk.) Stapf, *D. aristatum* (Foir.) C.E. Hubb, *Digitaria cruciata* (Nees ex Steudel) A. Camus, *D. setigera* Roth apud Roemer et Schultes, *Ehretia laevis* Roxb., *Euphorbia hirta* L., *E. pilosa* L., *Festuca kashmeriana* Stapf, *Ficus glomerata* Roxb., *F. neriifolia* Smith var. *nemoralis* Wallich ex Miq., *F. palmata* Forsk., *F. roxburghii* Wallich ex Miq., etc.

Plant species which contribute to decrease quantity and quality of milk are: *Andropogon echoides* L., *Artemisia nilagirica* var. *septentrionalis* (C.B. Clarke.) Pampanini, *A. parviflora* Roxb. ex D. Don,

Codonopsis ovata Benth., *Coriaria nepalensis* Wallich, *Erianthus munja* Roxb., *Erythrina variegata* L., *Fagopyrum cymosum* Treviranus, *Hedera nepalensis* K. Koch, *Hemarthria compressa* (L.f.) R. Br., *Hippophae salicifolia* D. Don, *Hypericum elodeoides* Choisy, *H. hookerianum* Wight and Arn., *H. nepaulense* Choisy, *H. oblongifolium* Choisy, *Impatiens scabrida* DC., *Myrsine semiserrata* Wallich, *Phacelurus speciosus* (Steudel) C.E. Hubbard, *Urtica dioica* L., etc.

Fodder Preservation

Preservation of high quality forages by drying is known as hay making. Hay making is the simplest way of fodder preservation and mountain farmers have been using this technique for centuries. In India sunlight is available plentifully for most of the days during the year. Farmers dry the green forage in sunlight in the open. This is the most economic method of preservation.

Hay making involves reduction of moisture content through sun-drying or artificial drying to 10-20 per cent which inhibits enzyme activity in the plants. Mountain farmers observe this scientific principle of hay making.

The harvested grass is preserved to be fed to livestock during winter season. People are well aware of the natural preservation methods, a technology which is centuries old. The grass is spread loose often on rooftops or on the broad fence around homestead and sun-dried for a couple of days. Moisture percentage is kept about 10-15 per cent. Also recommended by the animal scientists, this moisture level curbs microbial growth in the plants that could take place at appropriate moisture level, and thus helps preserve the fodder for longer duration.

Nutritive value of fodder is also taken care of in a scientific manner by the farmers when applying natural preservation technology. Leaves of the plants, which are more nutritive than the stem, are not allowed to shatter. Sun-drying also helps increase vitamin D content in the fodder.

The preserved fodder is stacked on the branches of trees or tied around a wooden pole near homestead. The stacking method is also a form of an appreciable art. The stacked fodder, locally called *luta* in Kumaun and *purolda* in Garhwal, is given the shape of a temple, broad base ending into a conical tip. This shape, 1.0 to 1.5m straight height, allows the rain water flow down the preserved material

Table 9.1: Indigenous Technology of Range Management in Uttarakhand Himalaya

Management Relating to	*Indigenous Technological Applications*	*Logic*
Site Specificity/Niche Ranges	1. Diversity at ecosystem level	1. Diversity of ecological niches according to altitude, slope direction, amount of precipitation, micro-climate, etc. is naturally available in the mountains (*e.g.*, Dune ranges, forest grazing lands, mid-altitude ranges, alpine pastures, trans-Himalayan ranges, etc.)
	2. Niche-specific vegetation	2. Niche-specific vegetation (*e.g.*, subtropical forages, temperate vegetation, alpine meadows, or sal type, oak type, alpine type, etc.) creates diversity amongst and within species that is used in range management and subsequently in nourishing livestock-dominated local livelihoods
	3. A mix of trees, shrubs and herbaceous plants.	3. A mix of vegetation types makes the rangeland ecosystems, more heterogeneous, ecologically more stable, more resilient and more productive.
Floral Composition	1. Periodic burning	1. Allows the herbaceous species (gramineae and non-gramineae families) to thrive on ground exposed to more solar radiation
	2. Cutting down and/ or thinning of forests.	2. Dominance of the vegetation of greater socio-economic importance
		3. More amount of fodder flow from the ecosystems for raising livestock

Contd...

Table 9.1–Contd...

Management Relating to	*Indigenous Technological Applications*	*Logic*
		4. Control of undesirable weeds through selective grazing;
		5. Maintenance of species of intangible uses contributing to vital ecosystem services
		6. Using plant-animal interactions for soil fertility management in croplands and for food production.
Cultivated Lands	1. Abandonment of agriculture on poor lands	1. Appropriate land use
	2. Practice of periodic fallowing	2. More efficient use of cultivated land
		3. Use of PPRs as CPRs that enhances socio-cultural cohesion within a community
		4. Improvement in the fertility status of soils of the fallow lands through preventing draining out of nutrients via food crops, allowing microbial degradation of crop residues (mostly roots), *in-situ* manuring by the grazing animals, and green manuring upon mixing up green plants/ weeds through ploughing at the end of fallow period.
Livestock Keeping	1. Native breeds of many species	1. Diverse species to take advantage of diverse agro-ecological zones
	2. Transhumance.	2. Livestock breeds with high feed conversion efficiency, adaptability and resistance against diseases, and capable of traveling long distances
		3. Transfer of energy and nutrients from more stable range ecosystems to more fragile croplands via livestock.

Contd...

Table 9.1–Contd...

Management Relating to	*Indigenous Technological Applications*	*Logic*
Ethno-vet Medicine	1. Use of various plants occurring in mountains' diverse ecological niches as medicine for livestock health management.	1. Ethno-medicines offer a cornucopia of benefits to the poor livestock-dependent communities as well as to the entire livestock sector
		2. Suitable for remote, inaccessibility-ridden poor areas inhabited largely by tribal and pastoral communities
		3. Cost-effective and safe approach to livestock health management
		4. No side-effects of medicine
		5. Availability in the vicinity of households.
Grazing	1. Rotational-cum deferred grazing of livestock	1. Avoidance of frequent defoliation and trampling of plants for quick regeneration of palatable plants
	2. Fencing	2. Plots left for harvesting grass for feeding during winter season
	3. Enclosure of areas/plots for harvest of grass at appropriate time for stall-feeding of livestock.	3. More appropriate use of uncultivated lands.
Distribution	1. Distribution of grass before harvest to all families on equal basis following social sanctions	1. Egalitarian distribution indicative of benefit sharing and social justice maintained through social sanctions
	2. Harvest by individual families.	2. Strengthening of local governance
		3. Appropriate participatory management of CPRs.

Contd...

Table 9.1–Contd...

Management Relating to	*Indigenous Technological Applications*	*Logic*
Harvesting	1. Annual cutting of grass after seed set and seed shed	1. Ecological regeneration of plants through vegetative propagation and germination of seeds shed in the grass plots
		2. Regeneration by seed shedding is taken advantage of so that root reserves of energy are not exhausted by frequent cuttings.
Livestock Feeding	1. Choosing tannin-free species with high crude protein and digestibility value;	1. Tannins are incriminating factors that decrease digestibility of crude protein;
	2. Response to seasonality and lean period	2. Livestock do not often face paucity of nutrients in their diet during scarcity period
	3. Preparation of concentrate incorporating locally produced food grains	4. The concentrate has high energy and protein value and is rich in overall digestible nutrients having bearing on the health, maintenance and productivity of livestock;
	4. Giving dietary allowance according to the physiological states of livestock	4. Dietary allowance takes care of the wear and tear of animal tissues and caters to the need of extra nutrient supplies during pregnancy, milk production and muscular work by the animals
	5. Preferring fodders that increase quantity and quality of milk.	5. Increasing quantity and quality of milk through selective/balanced feeding has implications on family's health and economy.

Contd...

Table 9.1–Contd...

Management Relating to	*Indigenous Technological Applications*	*Logic*
Fodder Preservation	1. Sun-drying of grass left loose for appropriate period to reduce moisture level to 10-15 per cent 2. Stacking of unchopped fodder in temple shape on tree branches or wooden poles.	1. Maintenance of nutritive value by keeping leaves intact with the plant 2. Preventing fodder from spoilage by microbial growth to help preserve the fodder for longer duration; 3. Increase of vitamin D content in fodder 4. Saving from the rain water by allowing minimum seepage.

quickly with minimum seepage, thus helping prevent the fodder from getting spoiled by increasing moisture content. In mountain areas there is no enough space that could be spared for storage of fodder. The traditional fodder stacking/ storage technique of farmers is an appropriate response.

Traditional knowledge relating to rangeland management in Uttarakhand Himalaya is summarized in Table 9.1. Traditional management (conservation and utilization) as being practiced in Uttarakhand and other parts of the HKH Region for millennia should not be overlooked in institutional development intervention. This could be instrumental towards evolving sound strategies of ecodevelopment in the fragile ecosystems.

Chapter 10
Natural and Anthropogenic Factors Operating on Mountain Rangelands

Rangelands are constantly subjected to certain kinds of factors that lead to certain sorts of stresses. These might be natural and anthropogenic factors. The natural ones could be categorized into abiotic and biotic factors, while the anthropogenic could be inherent with the livelihood systems and/ or these might be institutional factors, partly owing to policy framework under operation in the area.

Natural Factors

While a natural ecosystem, despite such stresses, tends to be in a state of ecological equilibrium, a managed one is bound to be severely constrained by such pressures.

Abiotic Factors

Abiotic factors are relating to all physical factors that operate on ecosystems that interact with the living beings and influence them in various ways. The abiotic factors operating on mountain rangelands, as on any other ecosystems, are the edaphic factors, the climatic factors, and the topographic or physiographic factors.

Edaphic Factors

Edaphic factors include the structure and composition of soil along with its physical and chemical characteristics. Soil, in fact, is a complex ecosystem, not merely a physical entity. Quality of soil and, consequently, the performance of a range ecosystem are influenced enormously by a mix of factors, like altitude, slope-orientation, precipitation, intensity and length of solar radiation, etc. There is an immense variation of soils in mountain areas. Different sites/ soils are unique habitats for unique fauna and flora and manifest into major types of vegetation and environment composition.

Soil is actually formed as a result of long-term process of complex interactions leading to the production of a mineral matrix in close association with interstitial organic matter, living as well as dead. In soil formation, modification of parent mineral matter takes fairly a long time. Such a modification is actually the result of interactions between climatic, topographic and biological effects (Sharma 2005). Soil thus can be called "soil complex" that has a biological system with mineral matter, organic matter or humus, soil water or soil solution, and soil atmosphere.

Each rangeland soil has distinctive flora as well as fauna of bacteria, fungi, algae, protozoa, rotifers, nematodes, oligochaetes, mollusks and arthropods. According to Docuchayev, a famous Russian pedologist of the Nineteenth Century, the soil is a result of the actions and reciprocal influences of parent rocks, climate, topography, plants, animals and age of the land. It can be represented by the following formula (cited by Sharma 2005):

$$S = (g \times e \times b.)\ \Delta t$$

where,

s: Soil; g: Geology; e: Environment; b: Biological influences; t: Time.

Climatic or Aerial Factors

The climatic or aerial factors include light, temperature of air (atmospheric temperature), rainfall or precipitation, humidity of air, and atmosphere (gases and wind). All these factors affect the performance of an ecosystem to a great extent.

Temperature shows pronounced temporal as well as spatial variations. In the Himalayan mountains of Kumaun and other areas marked seasonal extremes and fluctuations in temperature values is a normal routine. Altitude above mean sea level affects temperature values. In the Kumaun Himalayan mountains, we generally experience a decrease in temperature with increase in altitude. Thus, there is a different type of rangeland vegetation at different altitudes, showing distinct zonation from lower to higher altitudes. The successive zones of rangeland vegetation from base upwards are tropical and subtropical, temperate and alpine.

Rapid changes in the climate, which are indicative of the ominous greenhouse effect, remain one of the most talked about subjects and major concerns of humanity the world over. It is virtually causing all the components of nature to experience drastic changes. Greenhouse effect in the Himalayan mountains is increasingly becoming evident. Natural resource's becoming a victim of the ominous climate change is not an isolated phenomenon. Changing climate is resulting in the desiccation of the pastures. Lower water holding capacity of rangeland soils, decreased flow of moisture within an ecosystem, increased evapo-transpiration rate due to increase in temperatures, etc. are the major reasons of this desiccation, which affects the quantity and quality of biomass of the ecosystem.

Topographic or Physiogaraphic Factors

Mountains, hills, and hillocks are characterised by the inherent verticality. They owe verticality against gravitation. A slight disturbance in the highland ecosystems triggers mechanism of large-scale crumbling of the landmass. This problem is further aggravated by fragility factor. Mountain ecosystems are the most fragile and their crumbling as a result of physical and biological shocks is a matter of great concern (see also Figure 10.1).

There are two major physiographic features that determine the quality and performance of a rangeland ecosystem, namely, altitude from mean sea level and slope orientation (slope direction and

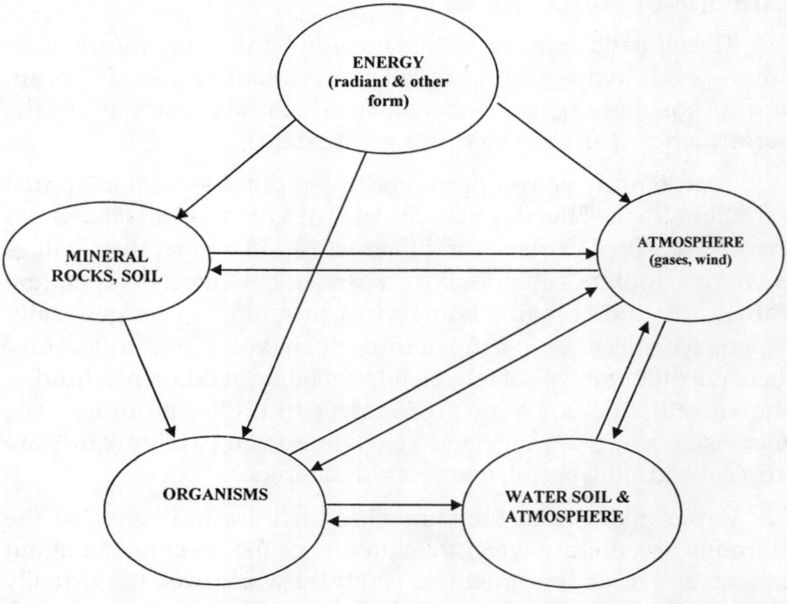

Figure 10.1: Interrelationships Between Principal Components of Environment and Rangeland Resources (after Sharma 2005)

steepness). With an increase in altitude above mean sea level, there are changes in temperature, pressure, wind velocity, and intensity of solar radiation. As a result, vegetation at different altitudes and slope aspects is different.

Steepness of a slope affects the amount of solar radiation received during the day and soil characteristics primarily through its effect on rate of water flow. South-oriented slopes receive solar radiation for longer duration during the day than the north-oriented slopes. The former types of slopes, therefore, are warmer than the latter. The other reason of comparatively higher temperature regime on south-oriented slopes is that they receive the rays of the mid-day sun almost at right angle, whereas the north-oriented slopes receive only oblique rays of the morning sun. Such local climatic differences create differences in the floristic composition of the rangelands. The northern slopes thus support virgin forests with hygrophilous ground vegetation while the southern slopes mostly have a sort of xerophytic vegetation. Floristic composition, despite these natural factors, however, is affected to a great extent due to constant anthropogenic activities in rangeland areas. Steepness of slope also determines the rate of water run off and soil erosion. Steeper slopes support little or no vegetation. Gentle slopes, on the other hand, harbour dense vegetation due to higher depth of supporting soil.

Even with heavy rainfall on a steep slope, active soil erosion and denudation due to run off result into disappearance of the plants from the area. Eroded hills, steep slopes, and rock faces may produce a special habitat, where only a particular group of plants can adapt themselves. Gravel, sand or clay eroded by water, wind, etc. from rock surfaces, are generally brought down by rivers and streams creating different habitats to various different kind of plants on certain sites (Sharma 2005). Exposure of slopes to strong winds also affects the type of vegetation in rangelands due to higher rates of soil erosion on such sites.

Biotic Factors

Biotic factors are all kinds of interactions between different forms of living organisms of a habitat–*i.e.*, plants, animals and microorganisms–which affect the ability of group of organisms to survive there. A number of biotic factors operate on rangeland ecosystems. However, the major impact is that of anthropogenic ones. Livestock are the major consumers of the producers thriving in the range habitat. Livestock consumption pattern, however, is managed by humans. Dimension of human interaction is so pivotal that it significantly alters type and composition and influences the functioning of all the three biotic components of an ecosystem, *viz.*, producers, consumers and decomposers.

Monocultures, particularly of *Pinus roxburghii* managed due to excessive human interference in the natural forests, are extremely vulnerable and liable to be damaged as a result of forest fires. Not only some 9000 km^2 area of the Uttarakhand Himalaya is covered with pine forest (Mehta 1996), this species is also rapidly encroaching on the realm of oaks, sal and other biologically mature forests. The fires cause incomprehensible damage and loss both to the forests and to the soil, which is deprived of its ability to infiltration of rainwater that recharges springs (Mehta 1996, Valdiya 1997).

Humans are the managers of the agro-ecosystems the rangelands are one of the components of. Through the grazing management and manual extraction of fodder, fuel wood, minor timber, bedding and mulch material, medicinal plants, and other non-timber products, man influences various populations of organisms–the biotic communities–and their environment. Domesticated animals–which are herbivores–are the major direct primary consumers consuming the foods synthesised by producers, the fodder-yielding plants. Herbivores that include a variety of insects, rodents, birds, wild grazing animals, etc. are other primary consumers. Secondary, tertiary and quaternary consumers also impart considerable impact on the composition and functioning of a rangeland ecosystem.

Decomposers meant to act upon the materials within the body of dead plants and animals are the invisible organisms of decay and play very important role within a range ecosystem rendering help in recycling the materials in the environment so that the process of life may go on and on like an unending chain.

Human role in managing a rangeland ecosystem paving way to maintain livestock-based livelihoods enormously affects species composition, populations, standing crops or the amount of living material at different trophic levels, food chain and food web; and with it, the very performance of the entire agroecosystem, the weather, the micro-climate and eventually the macro-climate. Impact of the natural factors operating on mountain rangelands is summarized in Table 10.1.

Anthropogenic Factors

There are several factors–physical, biological, socio-economic and institutional–operating on Himalayan rangelands. These factors concomitantly lead to the processes responsible for the degradation of the ecosystems. Degradation of these uncultivated core components of the mountain farming systems results in the decline of their own performance often indicated by declined biomass productivity. This state of the range ecosystems leads to reduced natural subsidies to the cultivated lands and livestock resulting in the decline in overall performance of the farming systems (Table 10.2).

Table 10.1: Natural Factors Causing Stress on Rangelands

Operating Factor	*Stress on Rangelands*
Abiotic Factors	
Edaphic Factors Structure, composition and physical and chemical characteristics of soils	1. Diversity of soils leading to low quality rangelands prevailing on large land tracts 2. Shallow soils with high percentage of gravels and rocks not supportive of high quality rangelands 3. Acidic soils with low quality, nutrient-deficient vegetation 4. Soils prone to erosion leading to nutrient paucity and resulted declined product 5. Low herbage productivity.
Climatic or Aerial Factors Light, air atmospheric temperature, rainfall, humidity of air, and atmosphere (gases and wind)	1. Distinctive zonation with some degraded rangelands 2. Changing climate resulting in the desiccation of the pastures; 3. Lower water holding capacity of rangeland soils; 4. Decreased flow of moisture within an ecosystem; 5. Increased evapo-transpiration rate due to increase in temperatures, etc.; 6. Poor quantity and quality of biomass of rangeland ecosystems.
Topographic or Physiographic Factors Altitude, slope direction, Slope steepness	1. Changes in temperature, pressure, wind velocity, and intensity of solar radiation create heterogeneity in floristic composition and ecological niches and adverse environmental conditions on regular/ seasonal basis 2. Presence of rocks and gravels in rangelands; 3. Desiccation of rangelands on south facing and steep slopes 4. Vulnerability to soil erosion, landslides, and biodiversity erosion 5. Continuous exposure of rangelands to environmental stresses due to operation of adverse factors.

Contd...

Table 10.1–Contd...

Operating Factor	*Stress on Rangelands*
Biotic Factors	
Interactions between living organisms and rangeland ecosystems	1. Excessive extraction of rangeland products, *e.g.*, fuel wood, fodder, grasses, etc. leading to a state of disclimax of rangeland ecosystems
	2. Trampling and browsing by livestock to the deterioration of rangelands
	3. Accelerated soil and biodiversity erosion and significant alteration in rangeland vegetative density and composition due to a complex of biotic pressures
	4. Impact on the functioning of all the three biotic components of an ecosystem, *viz.*, producers, consumers and decomposers.

Declined rangeland productivity as a result of operating factors, triggered processes and accentuating range degradation is revealed by several research findings. For instance, soil erosion rates on sloping lands have increased (20-30 per cent), part of agricultural lands has been abandoned due to decline in land fertility (3-11 per cent), appearance of rocks/ stones on cultivated lands has increased (130-200 per cent), forest area has declined (15-85 per cent), pasture/ rangeland area has declined (25-90 per cent), food grain production and self-sufficiency has reduced (30-60 per cent) (Shrestha 1992); rate of manure application has decreased (75 per cent), productivity of different crops has gone down (33-75 per cent), livestock holding size has reduced (63-70 per cent), livestock productivity for milk has decreased (50-67 per cent), time of fuel wood, fodder and bedding material has increased (50-80 per cent) (Singh 1992). Shengji (1996), Singh *et al.* (1998), Gaur (2004) and Singh and Gaur (2005) have indicated biodiversity erosion in the Himalayan areas.

These persistent negative indicators, indeed, emerge from the ecological degradation of what is the largest component of the mountains' geographical area, that is, the rangelands. The factors operating on range ecosystems generate processes that lead to eco-degradation of rangelands. The factors and the degradation processes are not separable. A single factor might trigger more than one processes and all factors often put the situation into a sort of ecological mess.

Table 10.2: Factors and Processes Leading to Rangeland Degradation

Factors	*Processes Leading to Range Degradation*
Physical Abiotic factors (Edaphic, climatic, and topographic), marginality, fragility, limited accessibility	1. Desiccation 2. Sheet erosion 3. Landslides 4. Acidic soils 5. Rocks and gravels 6. High degree of vulnerability
Biological Human populations, large number of livestock, exotic plants	1. Constant and excessive human pressure 2. High extraction rates 3. Overgrazing 4. Acceleration of soil and biodiversity erosion 5. Lower rangeland and livestock productivity 6. Transformation of rangelands into cultivated areas 7. Invasion of exotic species, such as *Parthenium, Lantana* and *Eupatorium* with concurrent loss of native species
Socio-economic Narrow land-man ratio, rangeland transformation into cultivated land, monocultures, migratory herds, tourism, construction activities, etc.	1. Decline in per capita rangeland area 2. Pressure for food production 3. Increasing ecological vulnerability 4. Increasing intrusion into and loss of virgin forests 5. Erosion of native biodiversity 6. Reduced application of manure in croplands 7. Increased soil erosion and landslides 8. Reduced biomass flows 9. Reduced agricultural and livestock productivity 10. Drudgery on women due to paucity of fodder, bedding material and fuel wood 11. Debasing of pastoralism and poor state of pastoralists
Institutional Poor understanding of rangelands, inappropriate pastoral policies, mismanagement (*e.g.*, growing monocultures for commercial gains), lack of appreciation for indigenous strategies, ignorance about rangeland contributions to environmental services, etc.	1. Declined flow of rangeland products such as wild fruits and vegetables, fodder, fibre, medicinal plants, etc. 2. Diminished generation of environmental services (like conservation of soil, water and biodiversity) 3. Erosion of natural biodiversity 4. Negative impacts on agriculture and consequent food insecurity 5. Reduced role of rangelands in maintaining people's livelihoods 6. Adverse impact on micro-climate

Chapter 11

Constraints in Range Management

Rangeland management in the Himalayan mountains is not as easy as it ought to be in the plains. It is because of the complex geographical, geological and ecological features of these fragile mountains. Added to these natural or inherent features are the problems created due to unique socioeconomic features and management and utilization patterns. Conventional institutional interventions have hardly resulted in something desirable for the effective range management–conservation and utilization. The present state of the range ecosystems leads to reduced natural subsidies to the cultivated lands and livestock resulting in the decline in overall performance of the farming systems.

Abiotic factors operating on Himalayan ranges include some of the inherent specificities of the mountain ecosystems. Emanating from the mountain perspective framework of Jodha (1990), these include limited accessibility, fragility and marginality. The other factors include those related to climate that become the major cause of what has been made concealed but is very potent: the desiccation of pastures.

Limited accessibility is one of the foremost features of mountain areas. Accessibility implies the movement of humans along with their domesticated animals to destined areas or pockets for the pure purpose of resource utilization. Accessibility is limited by the natural topography of the mountains. Rugged and treacherous terrain–further interspersed by steep slopes, deep gorges and streams and rivers–limits mobility of human beings and their livestock. Periodical seasonal hazards like torrential rains, hailstorms and heavy snowfall followed by the less controllable devastating landslides, avalanches, flashfloods, etc. further create conditions of inaccessibility, albeit temporarily, even in some areas which previously were well accessible. There are yet numerous areas in the Himalayan mountains that are left isolated and completely inaccessible.

While inaccessibility or limited accessibility is a constraining factor from the view point of resource utilization, it could be regarded as a positive attribute serving the great purpose of resource conservation the natural way. Inaccessibility has been overcome to a great extent by mechanical means such as road and bridge construction. Most of the rangelands in mountain areas are being increasingly accessible these days by mechanical means. The purpose, however, is not to provide facilities to livestock-dependent communities but to derive economic advantages by vested institutional or private interests to be met especially through expanding tourist and hotel industries extending into nature's virgin areas.

Himalayan mountains are the most fragile ecosystems on Earth. Fragility, a condition that results from the geologic, edaphic, and biotic factors associated high altitudes and steep slopes (Jodha 1990, Jodha *et al.*, 1992), is one of the major concerns of the socio-economic development of mountain areas. Owing to extreme degree of fragility, even the slight disturbances often trigger processes leading to vulnerability in the natural resources or ecosystems that might culminate into irreversible degradation. The natural relationship between deforestation, rain-drop impact and soil erosion, gullying and landslides, downstream impacts, and soil erosion imply equally serious consequences (Ives and Messerli 1989) and the fragility factors turn the whole situation grimmer.

Due to fragility factor inherently associated with the mountain mass, resource management in the area requires utmost care. Intensive use of the resources has to be avoided to avoid triggering mechanisms leading to a state of resource paucity and ecosystem degradation. Limited usage of the resources only lets the community compromise with the required biomass extraction. Mismanagement of range resources inevitably puts the whole system into a perilous state.

The term marginality often infers being outside of the mainstream processes that occur in the larger socioeconomic and political arena. In this sense, marginality is something that counts the least in the decision-making process (Jodha 1990). According to Singh and Tulachan (2002) there are three categories of marginality in mountain context, *viz.*, geopolitical, biophysical and socio-economic marginality. The term may also refer to the condition of biophysical resources including rangelands. Any resources that are too fragile, too unproductive, and too unattractive to be worthwhile for any use especially for those of the mainstream areas could be described as marginal. Both natural and manmade circumstances contribute to the marginal status of any resource of any community (Jodha 1990; Denholm 1991; Singh and Tulachan 2002).

Amidst the marginality-ridden mountain areas, rangelands are further marginalized due to their continued negligence. These ecosystems therefore have to bear the brunt of dual marginality. The marginality factor thus is one of the dominant constraints to the management of rangeland resources.

Misconceptions about Pastoralism

Negligence of range management emerges out of certain misconceptions. Some of the popular ones are:

1. Livestock are the cause of environmental degradation;
2. 'Overgrazing' is the major reason of environmental degradation;
3. Zero grazing (through adopting stall feeding, for example) is the precondition to regenerate forest areas;
4. Maintaining rangelands is uneconomic and these areas can be turned into more productive uses;

5. Large and unproductive herds are uneconomic and pastoralists are greatly responsible for ecological degradation in the mountains;
6. Pastoralists cannot contribute to rangeland improvement, etc.

Nature-livestock interactions are never negative (Tulachan and Neupane 1999). In reality, it is not livestock but the management systems that should be blamed for environmental degradation. It is now becoming increasingly apparent that existing paradigms for explaining the dynamics of rangeland ecosystems have not captured the dynamic nature of rangelands and, therefore, traditional measures for carrying capacity and range conditions have not been effective gauges for management in pastoral systems (Miller 1995).

Of late, grazing, often referred to as 'overgrazing', has been exaggerated by forest officials whose half-hearted efforts of plantation flopped like any thing. The issue of overgrazing was also used to 'justify' their efforts to plant unpalatable exotic plants. Natural advantages of moderate grazing are hardly appreciated.

In a region like the Himalayan mountains which is replete with rangelands and where pastoralism has been a natural way of living of livestock-dependent communities for millennia, the talk of zero grazing ensured by stall feeding is but insensitive.

It is but natural that, in order to maintain and enhance ecological stability, a large chunk of land should be covered with vegetation. Natural vegetation with high degree of biodiversity also contributes to enhance resilience of the whole ecosystem. Rangelands offer the vital services ensuring ecological integrity of the agro-ecosystems. Any other use of these ecosystems for example, terracing and cultivation of commercial crops, intensive tourism, construction works, etc. would prove extremely detrimental to the fragility and vulnerability-ridden mountains of the Himalaya.

Much of the land being used for grazing is marginal and unsuitable for cultivation, and ruminant livestock are the most efficient land use means to convert plants into protein (Miller 1995). Livestock for the pastoralists mean much more than they ought to be for others. Livestock for the planners might appear large in number and uneconomic, but for pastoralists they are manurers, insurance against losses, cushions against risks, efficient converters, source of their social prestige and part and parcel of local pastoralists' culture. Singh (2005) and Singh and Bohra (2006) have even attempted to establish that livestock contribute to the very essential ecological integrity of mountain areas, which is thanks to the vital role of livestock in nutrient flows in the agro-ecosystems.

Pastoralist/rangeland-dependent communities in the mountains have usually developed sophisticated ways of managing shared resources which outsiders often do not understand or acknowledge (Miller 1995). As has been discussed above in detail, farming communities in mountain areas have rich traditions of range management and their traditional range management strategies are not only time-tested but also futuristic.

Public institutes and policy makers might learn to appreciate farmers' traditional range management strategies. Rangelands mostly are in common property regime and their management stimulates social and cultural cohesion amongst village communities.

Popular myths and realities associated with pastoralism in mountain areas are summarized in Table 11.1.

The Least Understood Ecosystems

Rangeland ecosystems are the least understood ones. While a lot of work has focused on virtually all other ecosystems, *e.g.*, forests, deserts, lakes, rivers, oceans, etc., the rangeland ecosystems still face many issues lying unaddressed. The mainstream production systems in most of the plain areas are missing CPRs. In the mountains of the Himalaya, on the other hand, rangelands fall mostly in the CPR domain. Because of the negligence of rangelands in R&D system, several ecological and economic attributes of these ecosystems are not being articulated in the development policies and plans.

Erroneous Biodiversity Management

Vast areas under rangeland cover harbour valuable endemic biodiversity. The official management of these areas often ignores this biodiversity. Monocultures of commercially valuable species, such as of *Pinus roxburghii* and *Cedrus deodara* depict the common scenarios of rangelands under CPR regime. Such monocultures are extremely prone to the frequent incidents of forest fires which prove to be very devastating to the rangeland ecosystems. Monocultures infuse vulnerability in the whole system. Medicinal and aromatic plants, trees and shrubs providing highly nutritive fruits, wildlife, and other types of life forms, which could otherwise thrive in the rangeland ecosystems stocked with heterogeneity are deprived of chances to grow and flourish amidst monocultures.

Table 11.1: Myths and Realities about Pastoralism in the Himalaya

Myths About Pastoralism	*Realities About Pastoralism*
Livestock are the cause of environmental degradation	1. Nature-livestock interactions are positive 2. Livestock contribute to ecological integrity (through mediating nutrient flows) in an agro-ecosystem.
Overgrazing is the major reason of environmental degradation	1. Illicit felling, construction works, annexation for non-pastoral uses (*e.g.*, agriculture, tourist movement, etc.) are the major reasons of rangeland degradation 2. Overgrazing is hardly a problem in the transhumant pastoralism.
Zero grazing (through adopting stall feeding, for example) is the precondition to regenerate forest areas	1. Natural advantages of moderate grazing are accrued to nature as well as to livestock 2. Approach to stopping livestock grazing is insensitive, for it is the most favourable aspect associated with pastoralism.
Maintaining rangelands is uneconomic and these areas can be turned into more productive uses	1. A large chunk of land should be covered with vegetation for ecological stability/ integrity of agro-ecosystems and for ensuring goods and services on sustained basis 2. Any other use of the rangeland ecosystems would prove extremely detrimental to the fragility-ridden mountains.
Large and unproductive herds are uneconomic and pastoralists are greatly responsible for ecological degradation in the mountains	1. For pastoralists, livestock are manurers, insurance against losses, cushions against risks, efficient converters, source of social prestige and part and parcel of local pastoralists' culture 2. Valuation of livestock on account of their mediation in the provision of ecosystem services would prove them much more economic.
Pastoralists cannot contribute to rangeland improvement, etc.	1. Pastoralists' traditional range management strategies are not only innovative and time-tested but also futuristic 2. Public institutes and policy makers might learn to appreciate farmers' traditional range management strategies.

Rangeland development in the Himalaya often concentrates on livestock production rather than on multiple-use range resource management which provides for wildlife and the conservation of their habitat as well. Interactions between livestock and on the rangelands are still poorly understood and ways to minimize conflicts between livestock production and wildlife conservation still to be established (Miller 1995). A balance between natural vegetation, wild animals and livestock needs to be maintained which would ensure sustainability of the range ecosystems. This balance, unfortunately, is being felt in jeopardy which is the major constraining factor for the land-based economic development.

Inappropriate Pastoral Policies

Agriculture-related policies in the Himalayan mountains are not closely intertwined. Livestock and rangelands are grossly ignored in the R&D framework. Livestock development programmes are confined to only milk production mainly for supplies to the urban areas. For this purpose, crossbreeding of local cows with exotic bull semen is emphasized. Livestock role in the mediation of vital agro-ecosystem services is completely missing. Draught animal power (DAP) is of critical importance in the cultivation of mountain crops (Singh and Partap 2002), but it finds no mention in the policies. Cultivation of conventional fodder crops in the cropland area rather than the management and utilization of uncultivated fodder in the rangelands finds room in the policies. Mountain farmers, however, cannot spare any piece of their limited cultivated land for growing fodder. Issue of empowerment of rangeland-dependent communities is also missing.

The 'mainstream view', which maintains that pastoral practices need to be improved, has largely shaped pastoral development policy throughout the world (Sandford 1983). Traditional systems based on rangeland resources are very efficient, but policies in the Himalaya fail to appreciate this efficacy (Goldstein *et al.*, 1990; Miller 1995).

Rangeland science is not confined to merely livestock production and management, but now it has emerged as rangeland ecology. Rangeland ecology, therefore, needs to be developed like other branches of ecology, and like other disciplines of science. Himalayan rangelands offer ample opportunities for the scientists, who, with research focus on these neglected areas, can make significant contributions to the field of ecology as well as to the sustainable development of fragile mountains.

Summary

Rangelands comprise the largest geographical area in the Himalaya and constitute the core component of the mountain farming systems. These ecosystems not only render crucial ecological services, but also contribute to the livelihood security of livestock-dependent communities/ pastoralists dominating the region. Rangelands support a variety of plants many of which have good fodder value and are of great economic importance.

The present investigation is based on the extensive survey and collection of range plants from the Uttarakhand Himalaya. For the collection of plants and preparation of taxonomic accounts, frequent field trips were made during different seasons and months of the years. Several remote localities in the mid-altitude Himalayan mountains were covered for investigation. Vegetational analysis of grass-dominated rangelands was done to determine range community structure in terms of frequency, density, abundance and importance value indices of various species. Herbage productivity of different rangelands during different seasons was also determined. Utilisation pattern of different range resources was observed. Some fodder samples collected from the field were subjected to chemical analysis to determine their nutritive values and digestibility coefficients. Nutritive values of range fodders were also compared with crop residues, home-produced concentrate and commercial feeds. Rangeland contributions to agro-ecosystem integrity and sustainability were looked into and indigenous knowledge relating to range management was documented, analysed and interpreted. Various factors operating on rangeland ecosystems and constraints to range management were also analysed.

Diversity of the fodder plants rangeland ecosystems harbour is really amazing. About 550 species of grasses, other herbaceous plants, trees and shrubs were found in the sub-montane and montane Himalaya. The diversity of fodder plants, as enlisted after this exploration, is a proportion of the enormous biodiversity occurring in the parts of the Himalaya. All these plants virtually had certain fodder value. They are either grazed or browsed by livestock. Leaves of some woody species (trees and shrubs) are lopped for feeding of livestock at stall. Tender twigs, along with leaves of some shrubs and trees, are also consumed by livestock. Some plant species are grazed, browsed and eaten only by certain livestock species, not by all. In addition to the fodder value, many of the plants also provide fruit, fuel wood, fibre, flowers, dye, bee-forage, etc. Some species have unique ecological role, such as specific soil-binding properties, water conservation, pollination, etc. Some plants are of crucial ethnobotanical importance and are brought into various uses, such as in curing of certain diseases, amelioration of health, religious rituals, cultural rites, etc. Different parts (such as root, shoot, leaves, flowers, and seeds) of a species are used through certain physical and chemical processes in the curing of prevailing diseases, such as skin and eye problems, diarrhoea, dysentery, digestive problems, menstrual disorders, gonorrhoea, dysmenorrhoea, measles, respiratory problems, bronchitis, fever, malaria, dropsy, piles, diabetes, jaundice, suppressed urination, urinogenital problems, delivery-related complexities, etc. Many fodder species are of ethno-veterinary uses and people variously apply them in livestock health management. Certain species are regarded as sacred and are used in various cultural and religious rites. Local people in Uttarakhand mountains, as also in other parts of the Himalaya, are well equipped with the knowledge relating to the various uses of plants.

Farmers in the Himalayan mountains overwhelmingly depend on common property resources (CPRs) for livestock raising. Size of the holdings, therefore, is not a constraint to livestock husbandry in mountain areas as could be in the case of the plains where CPRs have become redundant or a thing of the past. CPRs are the major source of uncultivated green fodders. Private property resources (PPRs) provide bulk of the dry fodders in the form of crop residues. Tree/ shrub leaves and tender twigs, grasses from the field bunds and weeds extracted from the crop fields form greens for livestock available in the PPR-based land use. Green grass (hand-cut) accounted for 30 per cent of the total greens, while grazed grass as high as 55 per cent. Tree leaves amounted to only about 12 per cent. The other green comprising seasonal crops such as mustard, field weeds, etc. comprised about 3 per cent of the total green fodder given to livestock in a year, on an average. While virtually entire grazed grass was the product of the rangelands, most of the hand-cut grass (including weeds growing in the fields) came from the bunds of the terraced fields in the cultivated land area. Out of 12 per cent biomass of tree leaves, about 5 per cent came from the rangelands and the rest from the agroforestry systems.

Himalayan wildlife is unique and provides one of the most fascinating panoramas of biodiversity. Many of the wildlife species are endemic to the Himalaya. Wild mammals are adapted to high altitudes and some are confined to very small areas only. A number of wildlife species, due to several factors leading to considerable deterioration of their habitats in the Himalaya, are endangered/ vulnerable. There are some 69 mammals, 37 birds, three reptiles, and two amphibians that are endangered or vulnerable in the Himalaya. Rangelands are rich repositories of the fascinating wildlife and protection and conservation of these wildlife habitats is indispensable to avoid extinction of valuable and many endemic species in the Himalaya.

Seasonal availability of fodders is another important issue encountered in the area. *Quercus leucotrichophora, Ficus nemoralis, Morus serrata, Celtis australis, Ficus roxburghii, Ougeinia dalbergioides, Machilus duthiei*, etc. are the dominant tree species used for fodder lopping during summer season. *Ficus clavata, Bauhinia vahlii, B. variegata, Grewia optiva, Machilus duthiei, Quercus leucotrichophora, Ficus cunia* and *Vitis flexuosa* are the main fodder trees utilized during winter season. In rainy season, green grasses and other herbaceous plants comprise the major feed stuffs. While some green grasses are also available during summer season, in winter season rice straw, finger millet straw, barnyard millet straw and grass hay are the main dry fodders offered to livestock.

Interview with folk people in the study areas revealed that rangelands' first and the foremost utilization was for grazing the livestock. Fodder lopping was number two economic activity. Collection of bedding material to be used in the livestock sheds came at number 4. Agricultural tools and implements are made of certain woody species occurring in the rangelands, which was the fifth preferable use of these ecosystems. Collecting edible products for consumption purpose was the sixth preference of the people. At number seven came the use of rangeland-based species for mulch. Extraction of timber was the eighth preferable economic activity of the people. Medicine, dyes and edible fruits came next in order in that sequence for which people depend on rangelands.

A vegetational analysis of the herbaceous cover in the selected grass-dominated rangelands revealed that the *Themeda anathera* had the highest density of 72.5 per m^2, while *Cassia mimosoides* had the least density of 0.5 per m^2. *Themeda anathera* was the dominant grass in the rangelands of the study areas with highest importance value index (IVI) of 43 followed, by *Digitaria stricta, Pennisetum flaccidum, Setaria homonyma, Erigeron karvinskianus, Echinochloa crus-galli, Eleusine indica, Chrysopogon gryllus, Eragrostis atrovirens, Cyperus rotundus, Sporobolus diander, Oplismenus compositus, Arundinella nepalensis, Digitaria cruciata, Cynodon dactylon, Miscanthus nepalensis, Chenopodium ambrosioides, Agrostis pilosula, Cyperus nutans, Achyranthes bidentata, Pollinia quadrinervis, Dactyloctenium aegypticum, Justicia simplex, Brachypodium sylvaticum, Arthraxon prionodes, Leucas lanata, Oxalis corniculata, Rumex hastatus, Dichanthium annulatum* and *Cassia mimosoides.*

Average herbage productivity values in different rangelands–*viz.*, oak type, mixed type, oak-grass type, oak-conifer type and grass type–were 121.64±12.92g m^{-2}, 78.09±16.19g m^{-2}, 153.64±23.89g m^{-2}, 128.88±20.66g m^{-2} and 106.00±10.52g m^{-2}, respectively in summer season. In rainy season, productivity was more in comparison to that in the summer season, which was recorded to be 501.28±66.65g m^{-2}, 763.20±12.88g m^{-2}, 454.68±27.05g m^{-2}, 692.12±81.48g m^{-2} and 766.88±23.07g m^{-2} in oak type, mixed type, oak-grass type, oak-conifer type and grass type rangelands, respectively. Highest productivity was recorded in oak-grass type rangeland in summer season and from grass type rangeland in rainy season. In winter season, herbage productivity is critically low and ground flora, especially in grass type and conifer type rangelands, almost dries up. Herbage productivity was low in summer season and high in rainy season, which was due to the environmental factors which were unfavourable for plant growth during summer season.

Chemical composition based on proximate analysis and dry matter digestibility of some of the fodder species collected from households during study showed a wide variation in both within and amongst the types of fodder, *viz.*, tree leaves, grasses, crop residues and commercial feeds. The range of crude protein (CP), the most important indicator of nutritive value, was from 5.05 per cent (hay of *Themeda anathera*) to 25.02 per cent (*Grewia optiva*) in different groups. Among the tree leaves, the CP content was highest in *Grewia optiva* (25.02 per cent) and lowest in *Quercus leucotrichophora* (10.77 per cent). The CP content amongst grasses varied from 5.05 per cent (the hay of *Themeda anathera* grass) to 7.40 per cent (mixed green grasses). In crop residues, CP content was 3.23 per cent (rice straw) and 3.60 per cent (finger millet straw) and in commercial feeds CP content was highest in commercial feed urea-molasses-mineral blocks, *i.e.*, UMMB (22.14 per cent) and lowest in home-produced concentrate feed *Chaata* (16.19 per cent). Average CP content was 16.64±1.03 per cent in tree leaves, 6.32±0.43 per cent in grasses, and 3.14±0.185 per cent in crop residues and 18.32±1.91 per cent in commercial feeds. Among the groups, CP content was significantly different ($P<0.01$). There was significant difference ($P<0.01$) between tree leaves and grasses, tree leaves and crop residues, grasses and commercial feeds, crop residues and commercial feeds, but the difference was not significant between tree leaves and commercial feeds, grasses and crop residues.

Nylon bag dry matter digestibility (DMD) was significantly different ($P<0.01$) among the feed groups. The digestibility values ranged from 40.00 per cent (hay of *Themeda anathera*) to 90.00 per cent (*Chaata*). The average digestibility was 56.76±2.35 per cent in tree leaves, 47.40±2.69 per cent in grasses, 52.00±0.00 per cent in crop residues and 81.00± 6.65 per cent in commercial feeds. Digestibility was highest (67.00 per cent) in *Grewia optiva* and lowest (40.00 per cent) in *Ougeinia dalbergioides* in tree leaves. In grasses, it was the highest (55.00 per cent) in *Themeda anathera* green and the lowest (40.00 per cent) in the hay of *Themeda anathera.* In crop residues, it was the highest and the lowest in finger millet straw and rice straw, respectively. In the commercial feeds, the highest value (90.00 per cent) was recorded in *Chaata* and the lowest (68.00 per cent) in complete feed block (CFB). Digestibility was significantly different ($P<0.05$) in tree leaves and grass, tree leaves and commercial feeds, grass and commercial feed, crop residues and commercial feed ($P<0.01$), but not in grass and crop residues, tree leaves and crop residues.

Rangelands impart considerable ecological stability to the whole agro-ecosystem. They are also managed for sustainable production of useful biomass. Rangelands provide not only a variety of fodders for livestock, but also other useful biomass, such as wild fruits, uncultivated vegetables, mushrooms, edible flowers, buds and seeds, fibre, medicinal plants, honey, etc. Ecosystem services emanating from well-managed rangelands help maintaining an environment which is congenial for the productive services of a cropland. Conservation of soil and myriad life forms (biodiversity) in its own boundaries, nutrient and water/ moisture circulation in the whole farming system, maintenance of appropriate micro-climate, etc. are the other intangible attributes of the rangeland ecosystems in the mountains. Rangelands, thus, impart considerable ecological integrity and sustainability to the mountain agro-ecosystems.

Nutrients, a collection of chemical compounds, minerals and elements essential to the survival of living organisms, are taken up by the plants from the soil reserves and atmosphere. In a Himalayan agro-ecosystem, livestock play vital role in the nutrient cycling. Major chunk of the nutrients comes from uncultivated rangelands and transferred to cropland soils. Optimum amount of nutrients to flow into croplands is a precondition for the fertility management of the ecologically more fragile and more vulnerable system. In mountain agro-ecosystems, there is no considerable import of nutrients from outside the region. All the nutrients are produced from within the system itself. Moreover, livestock are not dependent on imported nutrients as would be the case in 'industrial' livestock production units. Animal-nature interactions in Himalayan mountains' smallholder systems indeed are environment-friendly. Livestock are the most important dynamic factor in managing the ecological or organic faming systems evolved in the mountains over millennia.

Management of the natural resources in the Himalayan mountains should be regarded not only essential but an imperative for restoring ecological balance and promoting livestock-based economic activities in the region. Rangeland ecosystems also serve as biological corridors making continuous or near continuous link to a suitable habitat through an inhospitable environment. Feed resources and appropriate habitat provided and ecosystem services rendered by the rangeland ecosystems are thus vital not only for livestock production, but also for livelihoods, ecosystem stability, sustainability and–consequently–for lasting peace and tranquility in the whole Himalayan region. Ecological balance of the agro-ecosystems maintained through efficient management of the rangeland ecosystems is the key factor to strengthen the rangeland-agriculture-livelihood linkages in the Himalayan mountains.

Range management appears to be the oldest art of natural resource management. Since the dawn of civilization, natural resource management systems have been the local community-based systems. Indigenous perceptions and practices regarding management of natural resources primarily concerned with protection and utilisation of biologically diverse resources. Natural resource management systems virtually are as diverse as the human-inhabited ecosystems on Earth.

As the world's largest mountain system, the Himalayan region has unique functions and roles and range management by local farmers and transhumant societies is the largest intervention into it. Development of silvi-pastoral systems conducive to regional geo-ecological specificities, livestock management, ethno-veterinary practices, grazing management, grass harvest, fodder distribution, livestock feeding management, fodder preservation and fodder stacking are some of the ways in which traditional management practices using local community-based knowledge are pivotal. Indigenous rangeland management in mountain areas should be properly understood, recognized and promoted for evolving ecologically sound, economically viable and socially just strategies of development in the fragile mountains.

Mountain rangelands are subjected to a number of stresses due to natural (abiotic and biotic) and anthropogenic factors. The factors operating on range ecosystems generate processes that lead to eco-degradation of rangelands. The factors and the degradation processes are not separable. One type of factor triggers more than one processes and all factors put the situation into a sort of ecological mess.

Several misconceptions are prevailing about Himalayan rangelands, for example, livestock are the cause of environmental degradation; 'overgrazing' is the major reason of environmental degradation; zero grazing (through adopting stall feeding, for example) is the precondition to regenerate forest areas; maintaining rangelands is uneconomic and these areas can be turned into more productive uses; large and unproductive herds are uneconomic and pastoralists are greatly responsible for ecological degradation in the mountains; pastoralists cannot contribute to rangeland improvement, etc. But these are far from realities. Rangelands, indeed, are the least understood ecosystems. Therefore, they are often missed in our development policy framework. Apart from the inherent mountain characteristics, such as limited accessibility, fragility and marginality, erroneous biodiversity management practices, and inappropriate policies happen to be the main constraints to range development in the mountains. Many of the socioeconomic and institutional constraints can be addressed effectively and appropriate perspective-based policies can be articulated in appropriate range management leading to phenomenal improvement in the supplies of goods and ecosystem services, livelihoods and living styles of mountain people.

Rangeland science now no more remains restricted to livestock production and management. It has assumed the shape of rangeland ecology. Himalayan rangelands offer ample opportunities for the scientists, who, with research focus on these neglected areas, can make significant contributions to the field of ecology as well as to the sustainable development of fragile mountains.

References

Agarwal, S.C. 1959. Grasslands of Chakrata Forest Division, District Dehradun. *Indian Forester*, 85: 659-662.

Ali, S. 1981. The Himalaya in the Indian ornithology. In Lall, J.S. (ed) *The Himalaya: Aspects of Change*. Oxford University Press, New Delhi.

Alirol, P. 1979. *Transhumant Animal Husbandry Systems in the Kalingchowk Region (Central Nepal): A Comprehensive Study of Animal Husbandry on the Southern Slopes of the Himalaya*. Swiss Association (SATA) for Technical Assistance, Integrated Hill Development Project, Kathmandu (mimeo).

Amatya, S.M. 1990. *Fodder Trees and Their Lopping Cycle in Nepal*. Technical Report, Kathmandu.

American Society of Range Management. 1964. *Glossary of Terms of Range Management. American Society of Range Management, USA*.

Anonymous. 1850. *Report of the Commissioner of the Patents for the Year 1849*, Part II. Agriculture. Washington, D.C.

Anonymous. 1986. Land resource mapping project. *Economic Report*.

Anonymous. 2001. *Genetic Diversity in Indian Trees.* Gene Campaign, New Delhi.

AOAC. 1995. *Official Methods of Analysis* (15th ed.). Association of Official Analytical Chemists, Washington.

Balaraman, N. 1981. *Feeds and Fodder Resources for Livestock in Sikkim.* Research Bulletin No. 13. ICAR Research Complex for NEH Region, Gangtok.

Bana, O.P.S. and Singh, V. 2004. Silvi-pasture systems for livestock production in the Hindu Kush-Himalayan region: issues and options. In Singh, V. and Gautam, P.L. (eds.) *Livestock Production Systems for Sustainable Food Security and Livelihoods in Mountain Areas.* GB Pant University of Agriculture and Technology, Pantnagar.

Bana, O.P.S., Singh, V. and Singh, A.K. 1996. Forage production in U.P. hills: status, constraints and development strategies. In *Proceeding of Sub-regional Working Group on Temperate Asia Pasture and Fodder Working Group (FAO) Meeting,* March 9-14, Dehradun.

Banskota, K. 2004. Rural livelihoods and agricultural biodiversity. *ICIMOD Newsletter,* No. 45.

Bhatt, A.B. and Rawat, N. 1993. Fodder resources of Garhwal: a search for non conventional fodder species. In Rajwar, G.S. (ed.) *Garhwal Himalaya: Ecology and Environment.* pp. 227-239. Ashish Publishing House, New Delhi.

Bhatt, P. 2005. *A Comparative Study on Carbon Pool in Distributed and Undistributed Studies of Oak (Quercus leucotrichophora).* M.Sc. (Environmental Science) Thesis. GB Pant University of Agriculture and Technology, Pantnagar.

Bohra, B. 2003. *Milk Production, Marketing and Consumption Scenarios at the Peri-urban Dairy Farms in the Champawat District of Uttarakhand.* M.Sc. Ag. Thesis. GB Pant University of Agriculture and Technology, Pantnagar.

Bohra, B. 2006. *Dairy Farming and Rangeland Resources in Mountain Agro-ecosystems in Uttaranchal.* Ph.D. Thesis. GB Pant University of Agriculture and Technology, Pantnagar.

Bohra, B. and Singh, V. 2005. Feed resources and utilisation pattern in the mixed mountain farming systems in Indian Central

Himalaya. In Rowlinson, P., Wachirapakorn, C., Pakdee, P. and Wanapat, M. (eds.) *Integrating Livestock-Crop Systems to Meet the Challenges of Globalisation*, Vol. 2. British Society of Animal Science, London.

Bourn, D., Maitima, J., Motsamai, B., Nicholson, C. and Sundstøl, F. 2005. Livestock and Environment. In Owen, E., Kitalyi, A., Jayasuriya, N., and Smith, T. (eds.) *Livestock and Wealth Creation: Improving the Husbandry of Animals kept by Resource-poor People in Developing Countries*. Nottingham University Press, U.K.

Brower, B.A. 1987. *Livestock and Landscape: The Serpa Pastorial System in Sangarmatha (Mt. Everest) National Park, Nepal*. Ph.D. Dissertation. Berkely: Department of Geography, University of California.

Burton, G.W. 1951. The adaptability and breeding of suitable grasses for the Southern United States. *Advan. in Agron.*, 3: 197-241.

Champion, H.G. 1936. A preliminary survey of the forest type of India and Burma. *Indian For. Rec.* 1: 1-286.

Champion, H.G. and Seth, S.K. 1968. *General Silviculture for India*. Government of India Publications, New Delhi.

Chatterjee, D. 1939. Studies on the endemic flora of India and Burma. *J. Royal Asiatic Soc. Bengal*, 5: 19-67.

Chaturvedi, V.B., Rasool, T.J. and Singh, V. 2004. Role of tree fodders in ruminant production in hill region: reconciling feed and environment security in the fragile ecosystems. In Singh, V. and Gautam, P.L. (eds.) *Livestock Production Systems for Sustainable Food Security and Livelihoods in Mountain Areas*. GB Pant University of Agriculture and Technology, Pantnagar.

Colby, W.G. 1941. Pasture culture in Massachusetts. *Mass., Agri. Exp. Sta., Bul.* 380.

Cronquist, A. 1981. *An Integrated System of Flowering Plants*. Columbia University Press, New York.

Dabadghao, P.M. and Shankarnarayan, K.A. 1973. *The Grass Cover of India*. ICAR, New Delhi.

Debnath, A. and Singh, V. 1990. *Bhimal* (*Grewia optiva*): wonder tree of the Himalaya. *Wasteland News*, Vol. V, May-July 1990.

Denholm, J. 1991. *Agroforestry in Mountain Areas of the Hindu Kush-Himalayan Region*. ICIMOD Occasional Paper No. 17. ICIMOD, Kathmandu.

Devarajan, S. 1999. *Effect of Tannins on the Ruminal Degradation Kinetics of Locally Available Tree Forages*. M.V.Sc. Thesis. Indian Veterinary Research Institute, Izatnagar.

Dugger, H.M. 1972. Foreword. In Younger, V.B. and McKell, C.M. (eds.) *The Biology and Utilization of Grasses*. Academic Press, New York and London.

Furgus, E.N. and Hollowell, E.A. 1960. Red clover. *Advan. Agron.*, 12: 365-436.

Gaur, R.D. 1982. Dynamics of vegetation in Garhwal Himalaya. In Paliwal, G.S. (ed.) *Vegetational Wealth of the Himalayas*, 12-25. Puja Publishers, Delhi.

Gaur, R.D. 1999. *Flora of the District Garhwal, North West Himalaya (with Ethnobotanical Notes)*. Transmedia, Srinagar, Garhwal.

Gaur, R.D. 2004. Vegetational diversity of Uttarakhand Himalaya and scope of regional development. In *Resource Appraisal, Technology Applications and Environmental Challenges in Central Himalaya*. HNB Garhwal University, Srinagar.

Gaur, R.D. and Maithani, D.D. 1990. Resource inventory and economic development with special reference to Garhwal Himalaya. In Maithani, D.D. (ed.) *Central Himalaya and Rural Development*, pp.55-61, New Delhi.

Gaur, R.D. and Tiwari, J.K. 1987. Some little known medicinal plants of Garhwal Himalaya: An ethnobotanical study. In: *Medicinal and Poisonous Plants of the Tropics* (ed. Leewenberg). pp.139-192. Netherlands.

Gaur, R.D., Bhatt, K.C. and Tiwari, J.K. 1992. An ethnobotanical study of Uttar Pradesh Himalaya in relation to veterinary medicines. *J. Indian Bot. Soc.*, 72: 139-144.

Gaur, R.D., Bisht, M.K. and Bhatt, K.C. 1988. A note on medicinal and other uses of some animal in Garhwal Himalaya: An ethnozoological perspective. *Man in India*, 68: 292-297.

Gaur, R.D., Purohit, V.P. and Silas, R.A. 1985. Land use pattern and settlement in Venu Valley, Rath area, Garhwal Himalaya. *Mountain Res. Dev.* 7: 91-94.

Gaur, R.D., Semwal, J.K. and Negi, K.S. 1983. Traditional fodder yielding plants of Garhwal Himalaya and their impact on quality and quantity of milk. *JOSHRAD*, 6: 89-93.

Gaur, R.D., Semwal, J.K. and Tiwari, J.K. 1984. A survey of high altitude medicinal plants of Garhwal Himalaya. *Bull. Med. Ethnobot. Res.*, 4(3-4): 102-116.

GB Pant University. 1982. *Integrated Natural and Human Resource Planning and Management in the Hills of UP*. GB Pant University, Pantnagar.

Goering, H.K. and Van Soest, P.J. 1970. Forage fibre analysis (apparatus, reagents, procedures and some applications). *Agriculture Handbook* No. 379: 1-20. Agriculture Research Service, United States Department of Agriculture.

Goldstein, M.C., Beall, C.M. and Chincotta, R.P. 1990. Indigenous nomadic pastoralism and ecological conservation on Tibet's 'Northern Plateau'. *National Geographic Research*, 6: 139-156.

Gupta, R.K. 1974. Synecology of the Grass Cover and Parallelism with Forest types in Garhwal Himalaya. *Indian Forester*, 100: 499-509.

Gupta, R.K. and Nanda, P.C. 1970. Grassland types and their Ecological Succession in the Western Himalaya. *Proc.11th International Grassland Congress*, pp.10-13.

Gupta, S.K. 1986. *Structure and Functioning of Natural and Modified Grassland Ecosystems of Garhwal Himalaya*. Final Technical Report Man and Biospheres, Department of Environment, Government of India, New Delhi.

Hanson, A.A. 1972. Breeding of grasses. In Younger, V.B. and McKell, C.M. (eds.) *The Biology and Utilization of Grasses*. pp. 36-51. Academic Press, New York and London.

Hutchinson, J. 1973. *The Families of Flowering Plants*. Clerendon Press, Oxford.

Ives, J.D. and Messerli, B. 1989. *The Himalayan Dilemma: Reconciling Development and Conservation*. pp. 295. Routledge, London and New York.

Jackson, M.G. 1985. A Strategy for improving productivity of livestock in the Hills of Uttar Pradesh. In Singh, J.S. (ed.) *Environmental Regeneration in Himalaya: Concepts and Strategies*. Gyanodaya Prakashan, Nainital.

Jackson, R. and Ahlborn, G. 1988. Snow leopards in Nepal: home range and movements. *National Geographic Research* 5 92): 161-171.

Jain, S.K. and Rao, R.R. 1977. *A Handbook of Field and Herbarium Methods*. Today and Tomorrow Print and Publ., New Delhi.

Jochim, M.A. 1981. *Strategies for Survival: Cultural Behaviour in an Ecological Context*. Academic Press, New York.

Jodha, N.S. 1990. *A Framework for Integrated Mountain Development*. MFS Discussion Paper No. 1. ICIMOD, Kathmandu.

Jodha, N.S. 1991. A Framework for integrated mountain development. In Singh, V. (ed) *Eco-Crisis in the Himalaya: Causes, Consequences and Way Out*. IBD, Dehradun.

Jodha, N.S. and Shrestha, S. 1990. *Some Conceptual Issues of Livestock Farming in the Mountains*. Mountain Farming Systems Discussion Paper No. 4. ICIMOD, Kathmandu.

Jodha, N.S., Banskota, M. and Partap, T. 1992. *Sustainable Mountain Agriculture*. Oxford and IBH Co., New Delhi.

Johnson, B.L. 1972. Polyploidy as a factor in the evolution and distribution of grasses. In Younger, V.B. and McKell, C.M. (eds.) *The Biology and Utilization of Grasses*. pp. 18-35. Academic Press, New York and London.

Joshi, M. and Singh, V. 2004. Ethnobotanical valuation of some plants in Himalayan Environments. In: *Environmental and Pollution Awareness in Hilly Areas*, Jyothi Prasad and K.N. Shukla (eds.), G.B. pant University of Agriculture and Technology, Pantnagar.

Joshi, M. and Singh, V. 2006. Ethnobotanical importance of some endemic plants in Central Himalayas. In Rawat, M.S.S. (ed.) *Resource Appraisal, Technology Application and Environmental Challenges in Central Himalaya*, pp. 192-199. Department of Geography, HNB Garhwal University, Srinagar (Garhwal).

Joshi, M., Bohra, B., Mandape, T., Joshi, S. and Singh, V. 2005. Ethno-veterinary practices in rural areas in India. In Rowlinson, P.,

Wachirapakorn, C., Pakdee, P. and Wanapat, M. (eds.) *Integrating Livestock-Crop Systems to Meet the Challenges of Globalisation*. British Society of Animal Science, U.K.

Joshi, N.K. and Singh, V. 2004. Natural resource management and rice landraces in the Himalayan mountains: some issues relevant for the International Year of Rice (IYR)-2004. *Environment and Pollution Awareness in Hilly Areas*. Pantnagar: GB Pant University of Agriculture and Technology.

Joshi, S. and Mandape, T. 2004. Hill Plants to the Rescue of Human Health: Evidence of Traditional Knowledge from Tehri Garhwal. In: Environmental and Pollution Awareness in Hilly Areas, Jyothi Prasad and K.N. Shukla (eds.), G.B. pant University of Agriculture and Technology, Pantnagar, India, 70-72

Kaul, R.N. and Ganguli, B.N. 1963. Fodder Potential of *Zizyphus* in the Shrub Grazing Lands of Arid Zones. *Indian Forester*, 39: 623-630.

Khan, M.H. 1979. A note on some grasses relished by marker in Chitral. *Pakistan Journal of Forestry*, 62-65.

Khatta, V.K, Tewatia, B.S. and Gupta, P.C. 1999. Nutritional evaluation of tree leaves of humid subtropical region of Himachal Pradesh. *Indian J Anim. Nut.*, 16(3): 220-223.

Khoshoo, T.N. 1995. Biodiversity in the Indian Himalayas: conservation and utilisation. In Shengji, P. (ed) *Banking on Biodiversity: Report on the Regional Consultation on Biodiversity Assessment in the Hindu Kush-Himalayas*. ICIMOD, Kathmandu.

Kundu, S.S. and Sharma, V. 1988. Chemical composition and *in vitro* dry matter digestibility of certain tree leaves. *Indian J. Anim. Nut.*, 5: 23-27.

Langer, R.H.M. 1963. Tillering in herbage grass. *Herb. Abstr.*, 33: 141-148.

Laude, H.M. 1972. External factors affecting tiller development. In Younger, V.B. and McKell, C.M. (eds.) *The Biology and Utilization of Grasses*. pp.146-154. Academic Press, New York and London, 146-154.

Lohan, O.P., Lall, D., Pal, R.N. and Negi, S.S. 1980. Cell wall constituents and *in vitro* dry matter digestibility of some fodder trees in Himachal Pradesh. *Forage Res.*, 6: 21-27.

Long, J.A. 1972. Developing superior turf varieties. In Younger, V.B. and McKell, C.M. (eds.) *The Biology and Utilization of Grasses*. pp. 53-65. Academic Press, New York and London.

Mandape, T., Joshi, S. and Singh, V. 2004. Use of ethno-phyto medicines in the hills. *Indian Farming*, 54(4): 22-23.

Meher-Homji, V.M. 1973. A phytogeographic approach to the problem of Mediterraniety in the Indo-Pakistan Subcontinent. *Feddes Repert*, 83:757-788.

Mehrez, A.Z. and Ørskov, E.R. 1977. A study of the artificial fibre technique for determining the digestibility of feeds in the rumen. *J. Agri. Sci.* 88: 645-650.

Mehta, J.S. 1996. Forest fires and land degradation in Uttarakhand. In: *Land Utilisation in the Central Himalaya*, (Ed. Kireet Kumar *et al.*), pp. 125-134. Indus Publishing Co., New Delhi.

Melkania, N.P. and Singh J.S. 1989. Ecology of Indian grasslands In: *Perspectives in Ecology*. (eds. J.S. Singh and B. Gopal) pp. 67-103. Jagmandar Book Agency, New Delhi.

Melkania, N.P. and Tondon, J.P. 1983. Grass cover and their biomass as influenced by protection in Western temperate Himalaya. *Forage Research*, 9: 181-183.

Miller, D.J. 1995. *Herds on the Move: Winds of Change among Pastoralists in the Himalaya and on the Tibetan Plateau*. MNR Series No. 95/2. ICIMOD, Kathmandu.

Mishra, R. 1968. *Ecology Workbook*. Oxford and IBH, Calcutta.

Mishra, R. 1983. Indian savannas. In: F. Bourliere (ed.) *Tropical Savannas*. pp.155-166. Elsevier, Amsterdam

Misri, B. 1995. Range and forest grazing in the Himalaya. In: Singh, P. (ed.) Workshop Proceedings on *Temperate Asia Pasture and Fodder Sub Regional Working Group*. pp. 28-33. Kathamandu.

Misri, B. 1999. The forage resource profile for India. In: Shelton H.M. (ed.). *Regional Research Centre Report*. pp. 1-12. Indian Grassland and Fodder Research Institute, HPKV Campus, Palampur (India).

Misri, R. 1986. Forage production in Kashmir Himalaya. In: Singh, P. (ed.) *Forage Production in India*. pp. 32-38. RMSI, IGERI, Jhansi.

Nautiyal, D.C. 1996. *Taxonomic Survey of the Poaceae from Garhwal Himalaya*. D. Phil. Thesis, HNB Garhwal University, Srinagar.

Negi, K.S and Gaur, R.D. 1994a. A contribution to the edible wild fruits of Uttar Pradesh Hills. *Bull. Bot. Surv. India*. 33:233-266.

Negi, K.S. and Gaur, R.D. 1994b. Principal wild food plants of Western Himalaya, Uttar Pradesh, India. In *Higher Plants of Indian Subcontinent*. 3:1-78. Bishen Singh Mahendrapal Singh, Dehradun.

Negi, K.S., Gaur, R.D. and Pant, K.C. 1995. Biodiversity: untapped plant wealth in Uttarakhand. *Indian Hort*. July-Sept., 32-42.

Negi, K.S., Tiwari, J.K. and Gaur, R.D. 1985. Economic importance of some common trees in Garhwal Himalaya: an ethnobotanical study. *Indian J Forestry*, 8: 276-279.

Numata, M. (ed.). 1983. Semi-natural pastures and their management in Eastern Nepal. In: *Structure and Dynamics of Vegetation in Eastern Nepal*. pp. 35-45. Laboratory of Ecology, Faculty of Science, Chiba University.

Odum, E.P. 1971. *Fundamentals of Ecology. 3rd edition*. Saunders, Philadelphia.

Oli, K.P. 2004. 'Corridors' in the management of biodiversity. *ICIMOD Newsletter*, No. 45, 13-14.

Oosting, H. J. 1958. *The Study of Plant Communities*. W. H. Freeman and Co, San Francisco.

Ostergren, G. 1949. *Luzula* and the mechanism of chromosome movements. *Hereditas* 35: 445-468.

Palmeri, R. 1976. *Domestication and Exploitation of Livestock in the Nepal Himalaya and Tibet: An Ecological, Functional and Cultural Historical Study of Yaks and Yak Hybrids in Society, Economy and Culture*. Ph.D. Dissertation. Department of Geography, University of California, Davis, CA.

Pandey, K.K. 1982. *Fodder Trees and Tree Fodder in Nepal*. Swiss Development Corporation, Berne.

Pariyar, D., Banstola, B.R. and Sedhain, G.K. 1996. *Fodder and Pasture Research and Development in Nepal. A Review and Synthesis of Its Application for Rehabilitating Degraded Land*, Lumle Centre, Lumle, Nepal.

Pathak, P.S. and Roy M.M. 1995. Agrosilvipastoral farming systems for optimizing forage and energy resources in the rainfed area. In: Singh, R.P. (ed.) *Forage Production and Utilization.* IGFRI, Jhansi.

Patil, B.D. and Pathak, P.S. 1980. Grassland development in the Himalaya. In: *Proceedings of National Seminar on Research. Development of Science and Technology*, Govt. of India, New Delhi.

Powell, M. and Ikpe, F. 1992. Fertilizer factories: nutrient recycling through livestock. *ILEIA Newsletter*, 8: 13-14.

Prieme, A. and Oksnebjerg, B. 1992. *Field Study in Shey Phoksundo National Park: Expedition Snow Leopard 1992.* World Wildlife Fund for Nature Report submitted to HMG Nepal Department of National Parks and Wildlife Conservation.

Puri, G.S. 1960. *Indian Forest Ecology*, Vol. I. Oxford Book and Stationery Co., New Delhi.

Ram, J., Arya, P. and Manral, J.P.S. 1993. Status and management of grazing lands in Central Himalaya, India. In Singh, V. (ed) Eco-crisis in the Himalaya: Causes, Consequences and Way Out. IBD, Dehradun.

Ranjitsinh, M.K. 1981. Himalayan fauna. In Lall, J.S. (ed) *The Himalaya: Aspects of Change.* Oxford University Press, New Delhi.

Raunkiaer, C. 1934. *The Life Form of Plants and Statistical Plant Geography: The Collected Papers of C. Raunkiaer.* Clarendon Press, Oxford.

Rawat, D.S., Rawat, A.S. and Gaur, R.D. 1993. Human activities and alpine plants: an impact assessment. In Rajwar, G.S. (ed) *Garhwal Himalaya: Ecology and Environment*, 107-130. Today and Tomorrow, New Delhi.

Reijntjes, C., Haverkort, B. and Water-Bayer, A. 1992. *Farming for the Future: An Introduction to Low-external-Input and Sustainable Agriculture.* Macmillan and ILEIA, Leusden, Netherlands.

Richard, C. (ed). 1994. *Natural Resource Use in Protected Areas of the High Himalaya.* IOF Project Technical Paper No. 94.

Richard, C., Basnet, K., Sah, J.P. and Raut, Y. (eds). 2000. *Grassland Ecology and Management in Protected Areas of Nepal.* Department of National Parks and Wildlife Conservation, ICIMOD and WWF, Kathmandu.

Roder, W., Wangdi, K., Gyamtsho, P. and Dorji, K. 2001. *Feeding the Herds: Improving Fodder Resources in Bhutan*. ICIMOD, Kathmandu.

Sandford, S. 1983. *Management of Pastoral Development in the Third World*. John Wiley and Sons, London.

Saxena, A.K. and Singh, J.S. 1980. Analysis of forest grazing land in part of Kumaun Himalaya. *Indian Journal of Range Management* 1: 13-32.

Schaller, G.B. 1977. *Mountain Monarchs: Wild Sheep and Goats of the Himalayas*. University of Chicago Press, Chicago.

Shankar, V. and Gupta, J.N. 1992. Restoration of degraded rangelands. In: J.S. Singh (ed.) *Restoration of Degraded Lands: Concepts and Strategies*. pp.115-155. Rastogi Publication, Meerut.

Sharma, D.D., Chawla, M.S. and Negi, S.S. 1966. Chemical composition and nutritive value of Bamboo (*Bambusa arundinacea*) and kachnar (*Bauhinia variegata*) tree leaves. *J. Res.*, 5: 438-442.

Sharma, P.D. 2005. *Ecology and Environment*. Rastogi Publications, Meerut.

Shengji, Pei. 1994. In Gurung, J.D. (ed.) *Indigenous Knowledge Systems and Biodiversity Management*. ICIMOD, Kathmandu.

Shengji, Pei. 1996. *Banking on Biodiversity: Report on the Regional Consultation on Biodiversity Assessment in the Hindu Kush-Himalaya*. ICIMOD, Kathmandu.

Shiva, V., Singh, V., Dankelman, I., Negi, B. and Singh, S. 2005. *Biodiversity, Gender and Technology in Mountain Agriculture: Glimpses of the Indian Central Himalaya*. Navdanya, New Delhi.

Shrestha, S. 1992. *Mountain Agriculture: Indicators of Unsustainability and Options for Reversal*. Mountain Farming Systems. Discussion Paper No. 32. ICIMOD, Kathmandu.

Sidahmed, A.E. and Rota, A. 2004. IFAD's Experience in Supporting Commmunity-based Livestock and Rangeland Development. In Richard, C. and Haffman, Kate (ed). *Strategic Innovations for Improving Pastoral Livelihoods in the Hindu-Kush-Himalayan Highlands, Vol. II: Technical Papers*. ICIMOD, Kathmandu.

Singh, H.J. 1989. Punjab's Problems of Plenty. *The Hindu Survey of Indian Agriculture,* Madras, 15.

Singh, J.S. and Gupta, S.R. 1993. Grasslands of Southern Asia. In: Coupland, R.T. (ed.) *Natural Grasslands: Eastern Hemisphere and Resume,* pp. 83-123. Elsevier Science Publishers, Amsterdam, The Netherlands.

Singh, J.S. and Saxena, A.K. 1978. The grass cover in the Himalayan Region. In: *Proc. National Seminar on Research, Development and Environment of the Himalayan Region,* pp. 164-203. Department of Science and Technology, Govt. of India, New Delhi.

Singh, J.S. and Singh, S.P. 1987. Forest vegetation of Central Himalaya. *Botanical Review,* 53: 82-192.

Singh, J.S. and Singh, S.P. 1992. *Forests of Himalaya.* Gyanodaya Prakashan, Nainital, India.

Singh, J.S. and Yadav, P.S. 1974. Seasonal variation in composition, plant biomass and net primary productivity of a tropical grassland in Kurukshetra. *India Ecol. Monographs,* 44: 351-376.

Singh, K.A. 1999. Nutrient contents in tree fodder and bamboo leaves of Eastern Himalaya. *Indian J. Anim. Nut.,* 16: 178-182.

Singh, P. and Misri, B. 1993. Rangeland resources: utilization and management in India. Paper Presented at *International Symposium on Grassland Resources* held at Huehot, Inner Mongolia, China, August 16-20, 1993.

Singh, R. 1997. *Economics of Livestock Production System in Himachal Pradesh* Agro-economic Research Centre, Himachal Pradesh University, Simla.

Singh, R. and Vaidya, C.S. 2002. Smallholder dairy farming in Himachal Pradesh, India: characteristics, constraints and development opportunities. In: Tulachan, P.M., Jabber, M.A. and Mohamed Saleem, M.A. (ed.) 2002. *Smallholder Dairy in Mixed Farming Systems of the Hindu Kush-Himalaya.* ICIMOD, Kathmandu.

Singh, R.P. and Hazra, C.R. 1995a. Forage seed production–status and strategy. In: Singh, R.P. (ed.). *Forage Production and Utilization.* pp. 309-323. IGFRI, Jhansi (India).

Singh, R.P. and Hazra, C.R. 1995b. Forage seed production perspectives and strategies. In Hazra, C.R. and Misri, B. (ed.)

New Vistas in Forage Production, pp. 323-334. AICRP (Forage Crops) IGFRI, Jhansi.

Singh, R.V. 1982. *Fodder Trees in India*. Oxford and IBH Publishing Company, New Delhi.

Singh, V. 1985. *Draoght Animal Power and Fodder Resources at Mid-altitude Himalayan Villages*. PhD Thesis. GB Pant University of Agriculture and Technology, Pantnagar.

Singh, V. 1989. Improving the productivity of forest/grassland-based farming systems in the UP Hills. *Advances in Forestry Research in India*, Vol. III, 157-176.

Singh, V. 1991. Strategy for increasing land and livestock productivity in the Himalayan environment. *ILEIA Newsletter*, Vol. 7, No. 1 and 2/ 91.

Singh, V. 1992. Indicators of unsustainability and contributing factors in mountain agriculture in Indian Himalaya. Paper presented at the International Workshop on *Approaches to Sustainable Mountain Agriculture* organized by ICIMOD at Manali, India from August 1 to 5, 1992.

Singh, V. 1993. Ecological farming systems for Garhwal. *Uttarakhand*, Vol. 7, 117-128.

Singh, V. 1994. Himalaya's vanishing heritage: oak. *Bhagirathi Ki Pukar*. Vol. IV, March 1994.

Singh, V. 1995. Biodiversity and farmers experiences from Garhwal Himalaya. Paper Presented at the *Beijer Research Seminar in Kota Kinabalu*, Sabah, Malaysia, 15-19 May, 1995.

Singh, V. 1996. Diversity in mountain agriculture. *ILEIA Newsletter*, April 1996

Singh, V. 1998a. *Drought Animal Power in Mountain Agriculture: A Study Perspectives and Issues in Central Himalaya, India*. International Centre for Integrated Mountain Development (ICIMOD), Kathmandu.

Singh, V. 1998b. Traditional biodiversity management strategies in the mountains: farmers' experimentation in Garhwal Himalaya. In: Singh, V. and Sharma M.L. (eds.) *Mountain Ecosystems: A Scenario of Unsustainability*. pp.25-66. Indus, New Delhi.

Singh, V. 1999. Trends and patterns of smallholder dairy farming in U.P. Hills. *Advances in Agricultural Research in India*, XI, 27-41.

Singh, V. 2000. *Smallholder Dairy Farming in Mixed Crop-Livestock Farming Systems in UP Himalaya*. ICIMOD, Kathmandu.

Singh, V. 2001. Sustainable development of smallholder dairy farming in Central Himalaya (Uttaranchal). Paper Presented at the Workshop on *Sustainable Development of Smallholder Dairy in Mixed Mountain Farming Systems of the Hindu Kush-Himalaya*, Dhulikhel, Nepal, April-18-20, 2001.

Singh, V. 2002a. Biodiversity: a key issue for commons management in the Himalayan Region. In Marothia, D.C. (ed.) *Institutionalizing Common Pool Resources*. pp. 519-530. Concept, New Delhi.

Singh, V. 2002b. Smallholder dairy farming in Uttarakhand, India: characteristics, constraints and development opportunities. In: Tulachan, P.M., Jabbar, M.A. and Mohamed Saleem, M.A. (ed.). *Smallholder dairy in Mixed farming systems of Hindu-Kush-Himalayan Regions*. pp. 53-70. ICIMOD, Kathmandu.

Singh, V. 2004a. Agrobiodiversity and food security in mountain areas: an Uttarakhand perspective. In: *Environment and Pollution Awareness in Hilly Areas*. Pantnagar: GB Pant University of Agriculture and Technology.

Singh, V. 2004b. Nature, biodiversity and food security: a glimpse of the Himalayan Mountains. In: *Souvenir: World Food Day–2004*. Pantnagar: GB Pant University of Agriculture and Technology.

Singh, V. 2005. Dairy development vis-à-vis agro-ecosystems in Himalayan Mountains: perspectives, issues and strategies in Uttarakhand, India. In Rowlinson, P., Wachirapakorn, C., Pakdee, P. and Wanapat, M. (eds.) *Integrating Livestock-Crop Systems to Meet the Challenges of Globalisation*, Vol. 2. British Society of Animal Science, London.

Singh, V. and Bohra, B. 2005. Livestock feed resources and feeding practices in hill farming systems: a review. *Indian Journal of Animal Sciences*, 75: 121-127.

Singh, V. and Bohra, B. 2006. *Dairy Farming in Mountain Areas*. Daya Publishers, New Delhi.

Singh, V. and Gaur, R.D. 2005. The Himalayan rangelands: ecosystem services and ecotourism opportunities. In Rajwar, G.S., Bisht, H., Sharma, Y.K., Kushwaha, M.D., Goswami, D.C. and Rawat, M.S. (eds.) *Tourism and Himalayan Biodiversity: Souvenir and Abstracts.* Govt. PG College, Uttarkashi.

Singh, V. and Kumar, A. 2002. Forests: source of fodder–an ecologically sound livestock feeding management in the Himalayan mountains. *Indian Farming* May 2002, 26-29.

Singh, V. and Naik, D.G. 1987. Fodder resources of Central Himalaya. In Pangtey, Y.P.S. and Joshi, S.C. (eds.) *Western Himalaya: Environment, Problems and Development,* Vol. I. pp. 223-241. Gyanodaya Prakashan, Nainital.

Singh, V. and Partap, T. 2002. Draught animal power in mountain agriculture: management scenarios in the Indian Central Himalaya. *Indian Journal of Animal Sciences,* 72: 1022-1033.

Singh, V. and Sharma, R.J. 1990. Forest-livestock-crop-human relationships and development of sustainable system: a Garhwal Himalayan case study. *Advances in Forestry Research in India,* Vol. V, 211-245.

Singh, V. and Tripathi, N. 2003. Foods in the wilderness: biodiversity in mountain forests holds key to food and environment security. *Indian Farming,* 57: 52-53.

Singh, V. and Tulachan, P.M. 2002. Marginal farming in mountain areas. *Asian Agri-History,* 6: 269-280.

Singh, V., Dankelman, I. and Solanki, A. 1998. Agroforestry–sustainability links in mountain context. *Advances in Forestry Research in India,* Vol. IXX, 1–17.

Singh, V., Dankelman, I. and Solanki, A. 1999. Common property resources in the Himalaya: biodiversity as the fundamental issue of development. *Advances in Forestry Research in India,* Vol. XX, 121–144.

Singh, V., Maurya, R.J. and Singh, Mahak. 2002. Agroforestry systems as a source of fodder for livestock in Uttaranchal. *Indian Farmers' Digest,* 34: 41-42.

Singh, V., Naik, D.G. and Kumar, A. 1988. Nutritional evaluation of fodder trees in Kumaun Himalaya. *Advances in Forestry Research in India.* Vol. II, 93-103.

Singh, V., Naik, D.G. and Kumar, A. 1990. Determining digestibility coefficients of fodder trees in Central Himalaya. *Advances in Forestry Research in India*, Vol. VI, 1-17.

Singh, V., Pandey, H., Singh, H. and Singh, M. 2003. Wildlife habitat and biodiversity in the Himalaya. In Thapliyal, D.C, Singh, J.L., Patel, M. and Singh, G.K. (eds.) *Holistic Approach to Wildlife Health Management and Conservation in Uttaranchal*. pp.10-20. GB Pant University of Agriculture and Technology, Pantnagar.

Singh, V., Sharma, M.L. and Singh, V. 1998. Himalayan forests in a state of unsustainability. In Singh, V. and Sharma M.L. (eds.) *Mountain Ecosystems: A Scenario of Unsustainability*. pp. 128-139. Indus, New Delhi.

Singh, V., Sharma, R.J. and Kumar, A. 1995. Perspectives on the utilization of forest fodder in the mountains. In Singh, V., Sharma, R.J., Kumar, A. (eds.) *Ecological Carnage in the Himalaya*. IBD, Dehradun.

Singh, V., Tulachan, P.M. and Partap, T. 2001a. Dynamics of dairy production in Uttaranchal hills. *Indian Journal of Animal Sciences*, 71: 1054-1057.

Singh, V., Tulachan, P.M. and Partap, T. 2001b. Livestock feeding management at smallholder dairy farms in Uttaranchal Hills. *Indian Journal of Animal Sciences*, 71: 1054-1057.

Singh, V., Tulachan, P.M. and Partap, T. 2001c. Smallholder dairy farms in the mixed mountain farming systems: a case of Uttaranchal Hills. *Indian Journal of Animal Sciences*, 71: 975-984.

Snedecor, G.W. and Cochran, W.G. 1968. *Statistical Methods*. The Indian Edition, Oxford and IBH Publishing Co., New Delhi.

Solanki, A., Sood, S. and Singh, V. 1999. Agro-horti-silvicultural systems in the mountains of the Hindu-Kush-Himalayan Region: A Framework for Development. *Advances in Forestry Research in India*, Vol. XX, 1-37.

Stebbins, G.L. 1972. The evolution of the grass family. In Younger, V.B. and McKell, C.M. (eds.) *The Biology and Utilization of Grasses*. pp. 1-17. Academic Press, New York and London.

Stoddart, L.A., Smith, A.D. and Box, T.W. 1975. *Range Management*. Mc Grara Hill Block Company, New Delhi.

Sundriyal, R.C. 1995. Grassland forage production and management in the Himalaya: A Review. *Journal of Hill Research*, 8: 135-150.

Sundriyal, R.C. and Joshi, A.P. 1991. Annual nutrient budget for an alpine grassland in the Garhwal Himalaya. *Journal of Vegetation Science* 2: 21-26.

Sundriyal, R.C., Chauhan, B.B., Kandwal. S.K. and Joshi, A.P. 1987. Vegetation composition of certain grasslands of Grahwal Himalaya as determined by soil profile and seasonal variations. *Indian Journal of Ecology*, 14: 37-46.

Takhtajan, A. 1981. *Flowering Plants Origin and Dispersal*. Smithsonian Inst. Press, Washington.

Tanaka, N. 1949. Chromosome studies in the genus *Carex*, with special reference to aneuploidy and polyploidy. *Cytologia* 15: 15-29.

Teli, B.L. 2006. Horticulture: economy and sustainable development in Uttaranchal. In Rawat, M.S.S. (ed.) *Resource Appraisal, Technology Application and Environmental Challenges in Central Himalaya*, pp.50-66. Department of Geography, HNB Garhwal University, Srinagar (Garhwal).

Thorne, R.F. 1968. Synopsis of a putatively phylogenetic classification of the flowering plants. *Aloso*, 6: 57-66.

Tiwari, C.B. and Dwivedi, V.K. 1997. Evaluation of shortfall in availability to requirement of feeds for livestock population in India. In National Symposium on *Feeding Strategies for Eco-friendly Animal Production in India*, Izatnagar. Feb 14-15, 1997.

Tripathi, H. and Mandape, M.K. 2002. Ethnoveterinary medical practices of Indian livestock owners in the northern plains of Uttar Pradesh. *Livestock International*, 6, 8-17.

Tripathi, R.K. 2001. *Energetics and Economics of Dairy Farms in Village Dairy Cooperatives in the Terai Area of Udham Singh Nagar*. M.Sc. Ag. Thesis. GB Pantnagar University of Agriculture and Technology, Pantnagar.

Tripathi, R.K., Tripathi, S., Selven, T., Kumar, K., Singh, K.K., Mehrotra, S. and Pushpangdan, P. 2004. Community structure and species diversity of Saddle Peak forests in Andaman Island. *Tropical Ecology*, 45: 241-250.

Tsuchida, K. 1983. *Grassland Vegetation and Succession in Eastern Nepal. In Structure and Dynamics of Vegetation in Eastern Nepal.*, ed. M. Numata, pp. 47-88. Laboratory of Ecology, Faculty of Science, Chiba University.

Tulachan, P.M. 1998. Livestock development in mixed crop farming systems: lessons and research priorities. *Issues in Mountain Development*, 98/5. ICIMOD, Kathmandu.

Tulachan, P.M. and Neupane, A. 1999. *Livestock in Mixed Farming Systems of the Hindu Kush-Himalaya: Trends and Sustainability.* ICIMOD, Kathmandu.

Tulachan, P.M., Jabbar, M.A. and Mohamed Saleem, M.A. (eds.). 2002. *Smallholder Dairy in Mixed Mountain Farming Systems of Hindu-Kush-Himalaya.* ICIMOD, Kathmandu.

Valdiya, K.S. 1997. Developing a paradise in peril. In: *Compendium of Pt. Govind Ballabh Pant Memorial Lectures.* pp.201-226. GB Pant Institute of Himalayan Environment and Development, Kosi-Katarmal, Almora.

Vallentine, J. 1971. *Range Improvement and Development.* Young University Press, Proso, Utah, USA.

Varma, A. 1997. Livestock: the great leveller of sustainable farming system. In: National Symposium on *Feeding Strategies for Eco-friendly Animal Production in India.* pp. 81-82. Izatnagar, Feb 14-15, 1997.

WCMC. 1992. *Global Biodiversity.* Champman and Hall, London.

Wegge, P. 1979. Aspects of the Population Ecology of Blue Sheep in Nepal. *J. Asian Ecology* 1: 10-20.

Whyte, R.O. 1974. Grasses and grasslands. In: *Natural Resources of Humid Tropical Asia.* pp. 239-262. UNESCO, Paris.

Wilson, P. 1981. Ecology and Habitat Utilisation of Blue Sheep *Pseudois nayaur* in Nepal. *Biological Conservation* 21: 54-74.

Yadav, P.S. and Singh, J.S. 1977. *Grassland Vegetation: Its Structure, Function, Utilisation and Management.* Today and Tomorrow's Printers and Publishers, New Delhi.

Yadava, A.K., Kumar, A. and Singh, V. 2003. Nutritive evaluation of some fodder plants in the hill and Tarai region of Uttaranchal. *Indian J. Anim. Sci.* 73: 793-797.

Zhaoli, Yan. 2004. Co-management of Rangelands: An Approach for Enhanced Livelihoods and Conservation. *ICIMOD Newsletter*, No. 45, 15-16.

Author Index

Subject Index

Index of Botanical Names

Index of Zoological Names